Water Regimes

In recent years the water sector has undergone profound institutional, economic and political transformations. Some countries have encouraged privatisation of water services, but in many cases this has provoked adverse reaction to such a neoliberal and market-based approach to this common shared but essential resource.

This book goes beyond the ideology of the public versus private water regime debate by focusing on the real policies implemented to provide better water services, particularly in urban settings. It asks the key question: what do the protagonists do? Numerous examples of alternative models are presented to show their social, institutional and technical-scientific characteristics and reveal who is responsible for implementing such systems. Policies are analysed in terms of their implications for employees and residents.

The book develops a new combinatory approach to water regimes, based on several international case studies from Argentina, Bolivia, China, France, Germany, India, South Africa and the USA, including a comparison of three cities in Africa. It is shown that the organisation of a water regime differs depending on the particular combination of institutions involved, and these combinations vary according to history and context. The case studies demonstrate the successes and problems of a range of private sector involvements in the provision of water services, and provide examples of how small-scale systems can complement large technical systems.

Dominique Lorrain is Director of Research (*emeritus*) at CNRS, Latts, Ecole des Ponts ParisTech, Paris, France.

Franck Poupeau is Director of Research at CNRS and Director of the International Joint Center iGLOBES, based at the University of Arizona, USA.

Earthscan Studies in Water Resource Management

Water, Power and Identity
The cultural politics of water in the Andes
Rutgerd Boelens

Water and Cities in Latin America
Challenges for sustainable development
Edited by Ismael Aguilar-Barajas, Jürgen Mahlknecht, Jonathan Kaledin and Marianne Kjellén

Catchment and River Basin Management
Integrating science and governance
Edited by Laurence Smith, Keith Porter, Kevin M. Hiscock, Mary Jane Porter and David Benson

Transboundary Water Management and the Climate Change Debate
Anton Earle, Ana Elisa Cascão, Stina Hansson, Anders Jägerskog, Ashok Swain and Joakim Öjendal

Rules, Norms and NGO Advocacy Strategies
Hydropower development on the Mekong River
Yumiko Yasuda

The Mekong
A socio-legal approach to river basin development
Ben Boer, Philip Hirsch, Fleur Johns, Ben Saul and Natalia Scurrah

Trans-jurisdictional Water Law and Governance
Edited by Janice Gray, Cameron Holley and Rosemary Rayfuse

Water and Rural Communities
Local politics, meaning and place
Lia Bryant with Jodie George

Land and Hydropolitics in the Nile River Basin
Challenges and new investments
Edited by Emil Sandström, Anders Jägerskog and Terje Oestigaard

Legal Frameworks for Transparency in Water Utilities Regulation
A comparative perspective
Mohamad Mova Al'Afghani

Water Regimes
Beyond the public and private sector debate
Edited by Dominique Lorrain and Franck Poupeau

For more information and to view forthcoming titles in this series, please visit the Routledge website: www.routledge.com/books/series/ECWRM/

Water Regimes

Beyond the public and private sector debate

Edited by Dominique Lorrain and Franck Poupeau

Routledge
Taylor & Francis Group
LONDON AND NEW YORK

earthscan
from Routledge

First published 2016 by Routledge

2 Park Square, Milton Park, Abingdon, Oxfordshire OX14 4RN
52 Vanderbilt Avenue, New York, NY 10017

Routledge is an imprint of the Taylor & Francis Group, an informa business

First issued in paperback 2019

British Library Cataloguing-in-Publication Data
A catalogue record for this book is available from the British Library

Library of Congress Cataloging-in-Publication Data
Names: Lorrain, Dominique, editor. | Poupeau, Franck, editor.
Title: Water regimes : beyond the public and private sector debate / edited by Dominique Lorrain and Franck Poupeau.
Description: Abingdon, Oxon ; New York, NY : Routledge, 2016. | Series: Earthscan studies in water resource management | Includes bibliographical references and index.
Identifiers: LCCN 2016011435| ISBN 9781138668201 (hbk) | ISBN 9781315618760 (ebk)
Subjects: LCSH: Water-supply. | Water utilities. | Municipal services.
Classification: LCC HD1691 .W329 2016 | DDC 363.6/1–dc23
LC record available at https://lccn.loc.gov/2016011435

ISBN: 978-1-138-66820-1 (hbk)
ISBN: 978-0-367-22763-0 (pbk)

Typeset in Bembo
by Wearset Ltd, Boldon, Tyne and Wear

Contents

Illustrations

Figure

Tables

Contributors

Julie Aubriot has a PhD in urban planning at the Ecole Nationale des Ponts et Chaussées (ENPC/LATTS), dedicated to "political uses of the right to water in poor neighborhoods of Soweto in South Africa". After an experience on cooperation programs in Senegal, she worked for four years at the Department of Water of the NGO Action contre la Faim (ACF) in Paris, and then at the World Water Council where she coordinated programs for the right to water at the sixth World Forum on Water. She is now a consultant on water issues at Hydroconseil, a consulting firm specializing in water and sewerage in countries of the Global South.

Rémi de Bercegol is a researcher at the Centre for Social Sciences and Humanities (CSH) in New Delhi and a member of the research unit LATTS (Laboratoire Techniques, Territoires, Sociétés – École Nationale des Ponts et Chaussées – Université Paris-Est). His research focuses on world urbanisation and the transformation of cities in the Global South, analysed principally in terms of their essential services (water, sanitation, waste management and energy).

Aymeric Blanc is an engineer from the Ecole Polytechnique and the Ecole Nationale des Ponts et Chaussées. He started work in 1997 at the French Ministère des Finances before integrating the Research Department of the Agence française pour le développement (AFD) in 2005 where he studies public–private partnerships.

Thomas Blanchet studied sociology at the University of Nantes and holds a PhD from the Freie Universität Berlin. He currently works as a research Fellow at the Nexus Institute for Cooperation Management and Interdisciplinary Research on a project dealing with citizens' participation in municipal water and energy utilities in France and Germany. He is also associate researcher at LATTS (Laboratoire Techniques, Territoires, Sociétés), where since 2014 he has worked on several projects on urban energy governance.

Sarah Botton is a sociologist associated with the LPED (UMR 151 IRD-Université de Provence, Marseille). She gained her PhD on water systems in Buenos Aires. She is now managing projects on water services and their funding for the Agence française de développement (AFD).

Murielle Coeurdray is a visiting research scholar in sociology and political science at the UMI/iGLOBES. Before joining the UMI, she worked in Berlin at the Centre Marc Bloch where she investigated the relationship between French water providers and German local councils in the aftermath of the fall of the Berlin wall. Her current research work is focused on providing a better understanding of the way environmental concerns arise and are being processed in the Western USA.

Bernard de Gouvello trained as an engineer and is a researcher at the Centre Science et Technique du Bâtiment (CSTB). He is associated with the Laboratoire Eau Environnement et Systèmes Urbains (LEESU) in the frame of a scientific partnership in which he coordinates projects on rainwater. He is also working on urban hydrology and sustainable water management.

Shankare Gowda holds a PhD in political science and is an independent researcher based in Delhi. His work focuses on exclusion processes and the analysis of political transformations in India. In the course of this research, he has conducted numerous field studies in various states of the Indian subcontinent.

Sébastien Hardy is senior researcher at the IRD and Director of the CEMCA Centre in Guatemala since 2014. He has worked in Bolivia for several years, realizing an *Atlas of Urban Risks* (IRD, 2012) in partnership with local administrations.

Dominique Lorrain is Director of Research (*emeritus*) at the CNRS, Latts, Chaire Ville, École des Ponts ParisTech. He is the author of several books on the management of urban waters, on mega-cities and more generally on the production of urban worlds.

Joan Cortinas Muñoz is a post-doctoral researcher at the UMI/IGlobes (CNRS/University of Arizona) Research Center in Tucson. He holds a PhD from the *Ecole des Hautes Etudes en Sciences Sociales* of Paris. His research focuses on public policy changes in two different fields: social policy and water supply policy. His investigation into social policy changes since the 1980s has been focused on Spain and France. His research on water supply policy in semi-arid regions is centred on the study of the west of the USA.

Brian O'Neill is studying water management at the Water Resources Research Center of the University of Arizona and he is associated with a collective inquiry on water conflicts in the Western USA in the frame of an ANR grant coordinated by the International Joint Unit iGLOBES (CNRS).

Franck Poupeau is Director of Research at the CNRS as well as Director of the joint international unit iGLOBES (Interdisciplinary and Global Environmental Studies) based at the University of Arizona. His work focuses on urban inequalities and water policies in the Americas.

Cécile Renouard is Professor of Philosophy at the Centre Sèvres and Director of the research program CODEV Entreprises et Développement des Pays Emergents of l'Institut IRENE (Institut de la Recherche et de l'Enseignement sur la Négociation en Europe) of the ESSEC.

Marie-Hélène Zérah is a research Fellow with the Institute of Research for Development (IRD). She has a PhD from the Paris Institute of Urban Studies. She was head of the urban dynamics research team in the Centre de Sciences Humaines of New Delhi between 2009 and 2013. She previously worked with the Water and Sanitation Program of the World Bank and the Suez Group. She was also involved in projects and consultancy for a number of organizations, including the European Union. She has worked extensively in the area of urban infrastructure, governance and democracy in Indian cities.

Acknowledgements

This book would not have been possible without the support of the Chaire Ville (Ecole des Ponts ParisTech) and its network of researchers and of the UMI iGLOBES (CNRS/University of Arizona). These partners and colleagues have been constructive in the discussion of the first versions of most of these chapters in French; these exchanges have contributed to enhance the current versions now published in English. The editors also wish to acknowledge the journal *Actes de la recherche en sciences sociales* for the evaluation of articles in their initial stage and for permission to publish in English several chapters included in this volume.

Introduction

What do the protagonists of the water sector do?

Dominique Lorrain and Franck Poupeau

During the 1990s, the water sector underwent profound institutional, economic and political transformations. Taking the view that infrastructure was an essential factor in development and that cities were strategic in the transformation of developing countries, international institutions have supported the idea of accelerating investment in urban infrastructure by inviting private sector firms on board (World Bank, 1994; Serageldin, 1994; Klein and Roger, 1994; Easter *et al.*, 1993; Israel, 1993; Kessides, 1993). The idea did not pose major problems in the transport, energy and telecommunications sectors. However, in the water sector things were more problematic (Guasch, 2004). It is true that, up until then, only the USA and France had elaborate institutional frameworks that made it possible to apply market principles to the management of this essential resource. But these reforms were sensitive for two reasons. At the time, the government of the United Kingdom was rolling out a vast programme of "privatization"[1] that was the polar opposite of the policy of the inter-war period; the radical nature of the approach was viewed by some as an archetypal example of the risks involved in property transfers. The period was also characterized by macro-economic policies that focused on aggregate outputs and a vision of the economy summed up in the "Washington Consensus" (Dezalay and Garth, 1998; Stiglitz, 2002). These policies were contested by an anti-globalist movement that often regarded the water sector as the prime example of a neoliberal agenda to dismantle public services. For these reasons, reforms in the water sector have often fuelled political passions.

However, a question remains. Beyond these controversies, what has really changed? This question is at the heart of this book. First, it may be imagined that such long-term disputes are not based exclusively on vague ideological preferences but are, rather, supported by legal rules and regulatory principles governing the sector. Second, the political over-determination in some cases should not obscure the real issue, namely that of providing "water for all"; consequently one should focus on the policies introduced in order to build, modernize and manage the socio-technical system: large-scale investment and ordinary policies. This descriptive approach prompts one to question the categories ordinarily applied to water services. Since the nineteenth century, the universal water services model has been presented as the most effective solution, but the difficulties in implementing this model in certain contexts – mostly poor developing countries – coupled with the emergence of alternative systems remind us that the very definition of the ideal model for managing water is still at issue.

The water service as an urban socio-technical system

Most social science researches on water begin by asserting its character as a resource that is indispensable to life. But, while water is a resource that has all the appearances of being natural and, therefore, resistant to sociological analysis, the fact that distribution networks have existed since ancient times (*Annales*, 2002; Scarborough *et al.*, 2003), and that, since the nineteenth century, such networks have played a vital role in the development of cities, means that water is even more likely to be taken for granted (Tarr and Konvitz, 1987; Tarr, 1989). In fact, like gas, electricity, transport and telecommunications, the water distribution and sanitation service is a socio-technical system that has become indispensable to the inhabitants of cities and large metropolises. Considering this sector as a technical system among others enables one to move beyond sector-based interpretations to which both the academic world and professional milieux tend to subscribe, thereby constituting a veritable "water community" (Meublat, 2001) with its reviews, its conferences, its international organizations and its national agencies. In effect, the water sector generates a vast field of applied research in which public institutions, the R&D departments of private corporations and consulting firms are all involved – exchanging ideas, good practices and evaluations. These studies on the water sector over several decades have been conducted to elaborate a *doxa* that contributes to the development of both consensus among professionals and public opinion that finally influences the choice of political decision-makers (Payen, 2013: 119).

In regard to developing countries, one of the most accepted ideas of this *doxa* is that water management is too costly for poor populations which, by definition, have low incomes. But convergent studies conducted in countries of the "South" show that the cost of water for poor populations with no access to the public service (and who are forced to purchase water in bottles or water tanks, or to finance their own rainwater storage equipment) is higher than for people who are connected to it (Zérah, 2000; Brook and Irwin, 2003; Harris, 2003: 27; Jaglin, 2005; Franceys and Gerlach, 2008). Furthermore, it is considered that operating a network in a sector characterized by a low level of profitability,[2] without substantial public investment, notably in terms of infrastructure in the poorest neighbourhoods, may appear to be an arduous undertaking. Nevertheless, the comparison with – among other technical systems – cellphone networks, which have developed very rapidly over the course of the past 15 years, seems to suggest that, even in the absence of public subsidies, poor populations have the means to pay for services adapted to their financial circumstances. This is one of the paradoxes inherent in political approaches to urban services. As recently demonstrated in Bolivia, users can prefer approaches that focus on delivering improved conditions of access rather than a free service (Poupeau, 2011b). The key question is not only one of cost; it is also the service which is provided to people as well as their desire to have it, and behind that the allocation of revenue among different forms of consumption.

Another characteristic of this *doxa* is its focus on the public or private status of the operator. This feature has been particularly evident since the 1990s when the generalization of the universal service model achieved international recognition. During this period many contracts were signed but numerous ones were characterized by disputes over the contracting-out of the service (Guasch, 2004) in cities including, but not limited to, Buenos Aires, Berlin, Cochabamba, Le Paz-El Alto and Soweto. By 2010, private water companies had withdrawn from many developing countries. Nevertheless, in these problematic cases the number of connections generally increased (Marin, 2009), whereas many cases of remunicipalization failed to produce the kinds of economic and social results that had originally been hoped for. Cochabamba, emblematic of the "Water Wars" of the early 2000s, provides us with a stark example.

Another commonly shared idea is that providing the poorest populations with access to basic services depends on the promotion of a "right to water" (Bakker, 2010). Of course, this right may be used to support a "communitarian management" structure capable of taking into account criteria of justice and democratic participation. But, at the same time, while asserting a principle, it does not guarantee an improvement in the service. In certain contexts, its application, which is by no means guaranteed, can lead to stocks of the resource being exhausted. Similarly, the notion according to which a call for consumers or stakeholders to take part in the process would be sufficient to render the services more efficient is far from having been validated. Above all, the concept of the "right to water" does not tell us whether a universal extension of services is the only way forward, the most suitable approach to confronting the issues of urban sprawl and the growing scarcity of natural resources characteristic of regions marked by "water stress", including the American West, the high plateaux of the Andes, the *Sertão* in Brazil, North Africa, the Arabian Peninsula, southern India, central Asia and northern China.

This water technical system is embedded in the materiality of social life as well as locality; this system possesses a morphology and a series of properties (availability of resources, rainfall, relief, ecosystems) from which both economic (fixed assets) and legal data (contracts) may be derived. It is characterized as well by an institutional framework that makes it possible to manage the water system in a given territory marked by a particular history and occupied by social groups with their specific interests and conflicts of interest. These are the factors on which we have based our hypothesis, choosing to focus on what protagonists actually do, rather than on the public or private status of the operator.

The operators, the technical networks and their practices

The characteristics of water as a socio-technical system, and the profusion of views concerning the legitimate and the "best management" structure that should be applied to it, encourage us to examine *what the protagonists of the sector do*, rather than what they say they do, or what they should do, and to analyse practices in their social contexts. Our approach is to bring the socio-technical system to the centre of the analysis in order to avoid an imprecise use of the notion of "actor" as is too frequently the case when it is applied to any collective action. The notion of "actor" becomes abstract, dissolving the practices in an absence of social characteristics. There is, therefore, a risk of viewing the system of action as more important than the action itself. Furthermore, this kind of blurred semantics hides presuppositions that are not discussed and generates an exclusive interpretation of negotiation-based public policy inspired by game theory. This approach, based on the notion of "games between actors", presupposes that everyone shares the objective of a satisfactory result. In fact, it is not only a game: behind the tap there is a complex infrastructure made up of pipes, pumping stations, sub-stations, tanks, sensors, meters and so on. This represents huge fixed assets, and contributes to the organization and structure of cities. If any improvement in the quality of service is obtained it is first because investments have been made in either the construction of large equipment or regular maintenance works. The perspective developed here consists of focusing on *results*, defined not exclusively from the point of view of economic profitability but also from that of approaches towards implementing socio-technical systems. Who is responsible for implementing such systems? What are their social, institutional and techno-scientific characteristics? What policies do they apply to the network in regard to employees and residents? And to what degree does the involvement of the latter help improve the service?

On this point, it is useful to recall the kinds of ambitions underpinning the development of urban technical systems in the second half of the nineteenth century. At that time, the question of means was secondary and the debate concentrated on how to bring "progress" for all, first, by putting an end to the typhoid and cholera epidemics that caused huge death-tolls (Goubert, 1986: 46; Lequin, 1983: 277), and second, by providing homes with tap water. In a relatively short space of time, several urban infrastructures, including the distribution of drinking water and the evacuation of wastewater as well as electricity, gas and light, were introduced in a number of cities, resulting in a huge improvement in the quality of urban life (Hughes, 1983; Tarr, 1989). This brief historical review encourages the exploration of another perspective. It suggests that we should pay less attention to the legal identity of the operator and how it is selected and, instead, focus on how a process of modernization is engaged; this invites an examination of what the operator does, what industrial policies it applies and the results it produces. The hypothesis that emerges from these considerations is that the quality of operators cannot be defined on an a priori basis: there is no clear-cut dichotomy between, on the one hand, innovative private companies and, on the other, inefficient public bureaucracies, any more than there is a divide between profit-hungry multinationals and naturally virtuous municipal corporations devoted exclusively to the social benefit of their actions.

There is another avenue to be explored, with the focus on how water resources are effectively used. In effect, inequalities in access are less dependent on disparities in natural resources and more dependent on difficulties in implementing services in areas inhabited by the poorest populations: see the case of equatorial countries in Africa and Asia that suffer from poor services in a situation of permanent and abundant rainfall. This is a good reason for taking an interest in the policies implemented by operators in these contexts. In this perspective, since Douglas North (1990) and, above all, Elinor Ostrom (1990), the neo-institutional literature has focused on the organizational dimensions of managing services. Taking issue with the notion of the "tragedy of the commons" (Hardin, 1968; Ingold, 2008), this demonstrates that water shortages are not ineluctable events linked to the disproportionate growth of populations faced with limited natural resources, but are, instead, often the result of institutional shortcomings. In many situations, water as a "service" is, for management purposes, separated from water considered as a "resource" (on a large scale). In such contexts, it is thought of as "an artifact produced by the skills of engineers, most frequently flowing through pipes buried in the ground, or, in other words, separated from nature" (Barraqué, 2011: 4).

Questions of method

At first glance, the application of a monographic approach, based on narratives concerning the protagonists of the water sector, may seem to be the most appropriate method. But by basing such work essentially on interviews, we run the risk of failing to capture the big picture. The interview-based method not only tends to reveal salient moments, including disputes and conflicts emerging in the public space, but it also occludes the often invisible constraints faced by socio-technical systems, small, ordinary operations that largely explain the efficiency of water distribution and sanitation systems and their capacity to adapt to needs. This general problem becomes more pronounced in sensitive issues such as the private status of the operator. Interview-based surveys run the risk of focusing exclusively on the perceptions of visible "actors". In addition to the fact that it is sometimes difficult to express the diversity of points of view, the interviewees themselves have, by definition, selective memories. Meanwhile, operators often account for only part of the story, as a consequence of which some periods are overrepresented, while others are characterized by a deficit of information.

One of the essential methodological rules governing field surveys is that a number of different sources of information must be taken into account, including contractual documents, statistics, articles in daily newspapers that serve as a chronological marker, official reports, studies sponsored by specific bodies, municipal and national databases, the comments of interviewees, and information collected in questionnaires. Taken individually, none of these sources encompasses the system as a whole, but an approach based on methodological pluralism makes it possible to check the data and reveal something that encourages us to look beyond first impressions. In effect, the arguments applied are often linked to technical data that need to be verified. One thus has to take into account factors such as the nature of the contract in question, the history of the technical system, its implantation of the metropolitan space, socio-spatial inequalities affecting household incomes, and access to essential services.

The comments of interviewees must be accompanied by data from outside the operator so that its practices may be situated in a chronological context (a recourse to history) and located within more all-encompassing global cycles (political economy, the international circulation of management models). When the operator belongs to a private multinational group, the constraints to which that group is subject must be considered, which implies that an analysis based exclusively on local conditions would be distorted. Similarly, when tariffs are raised, one need to examine factors other than the percentage increase, and also take into account the monthly bill per household which, when compared to income, reveals its impact upon household expenditure as a whole. An analysis of tariffs is, in fact, linked to the more general problem of the profitability of the water sector and of the way in which it is calculated: should one use the profit/turnover operating ratio employed in the financial press as a rustic instrument of comparison, or the Chinese method (a percentage of net assets), or the "cost-plus" method (capital cost plus a certain percentage)?

In summary, a multi-level perspective is required in order to take into account all factors involved in a socio-technical system. A successful analysis must encompass the issues faced by socio-technical systems, systems that are both the subject of contracts and the basis of the service. Sociological approaches to these systems should explore the practices, interests and conflicts of the various protagonists by applying an analytical frame covering the constraints, rules and norms governing the implementation of those systems. All the cases presented in this issue are based on an ensemble of common variables, including the initial "problem" to be solved (in terms of implementing a service), the choice of the operator and its partners (with their histories and interests), state intervention and the industrial policies developed, the kinds of management instruments employed, and the values and beliefs on which operators – public or private – base their legitimacy.

This insistence on the "problems" faced by socio-technical systems does not imply a technological determinism that limits choices and margins of action. On the contrary, the variety of rules and norms (and in general of "small" institutions) provide resources that may be used by various protagonists to impose or contest a given water regime. The water sector is a kind of intellectual battlefield, not only because the disputes by which it is characterized focus on effective management models – and the technical procedures on which they are based – but also because they address the problem of the definition of those issues and, thus, imply competition between different models: public versus private, large technical systems versus decentralized solutions.

Debates on major contracts in the water sector illustrate the *undecidability* of certain social phenomena. Due to their complexity, contracts may have different interpretations that are neither entirely wrong nor entirely right because their interpretation depends on contextual factors. In the case of Buenos Aires a specific programme designed to bring the water distribution service to poor neighbourhoods in partnership with the residents suggested that water management is "within everyone's grasp"; as a consequence the legitimacy of the enterprise, based on its technical competence, was undermined. However, if the programme demanded by local politicians had not been implemented, the company would have been guilty, first, for not being reactive, and second, of failing to take an interest in the least profitable sectors of the concession. The same situation may thus be interpreted in different ways. The operator can either be seen as reactive and pragmatic, and willing to seek innovative solutions, or as delivering a mediocre service to people living in poor neighbourhoods in order to protect its operating profit.

Similarly, the contract between the Ville de Paris and the Compagnie générale des eaux incorporated certain principles outlined in the 1860s. But one should not interpret practices applied in the past via the prism of contemporary categories. Indeed, the principles underlying nineteenth-century contracts are very different to those to be found today in the theory of agency and the separation between the principal (the public collectivity) and the agent (the delegating enterprise). A militant interpretation may highlight the fact that the private operator exploited informational asymmetry, behaved opportunistically, and did everything needed to ensure that it had the political class in its pocket. Another may emphasize progress made during the 1990s/2000s in terms of transparency and the formalization of contractual obligations, with the growing "juridicization" of the water sector. A sociological interpretation may thus consist in showing that the contract reflected the state of institutional relations in France which, until recently, were characterized by weak local authorities that relied on the support of large companies and the state (Lorrain, 1992). This trio of organizations constitutes a politico-industrial system in which public and private elites, educated in the same *Grandes écoles*, and who, therefore, share a specific idea of development, take the most important decisions based on the same knowledge tools. This system is by no means limited to the water sector. The same decision-making mechanisms are to be found in policies governing transport, telecommunications and the nuclear industry. In other words, interpretation depends on perspective. If the water sector is considered to be autonomous, it is possible to assert that old practices could be subject to criticism because they were neither transparent nor balanced enough to meet current standards. But if the past is taken into account, considering the logics of practices in the society of a given era, systems respond to the categories of their time. Therefore, a descriptive approach is necessary to explain how they actually function.

Organization of the book

The objective to go beyond the debate between public and private sectors relies on the statement that the legal status of companies does not report the outcomes of water utilities or their industrial policies. The combinatory approach developed in this book consists in analysing the impact of several determinant factors such as the nature of the contract, the implementation of local partnerships, the consequence of culture and values, the scale of prices and tariffs or the structure of urban inequalities against a background of case studies allowing them to experiment with various impacts.

Part I ("Unorthodox models and practices") explores the diversity of combinations between public and private sectors, based on two exemplary case studies in China and United States. In Chapter 1, Dominique Lorrain focuses on a joint corporation operating in the field of water services created by the municipality of Chongqing and the group Suez in China. The joint enterprise model deserves close scrutiny even though it is heterodox in the frame of the dominant organizational theories of monopolistic industrial networks. Its main principle is neither the separation between a principal and an agent nor a detailed contract, but co-production. The partners involved cooperate at all levels of production: shareholders of the exploitation company; sharing of executive instances; and network exploitation. This solution was developed in China. This chapter uses a descriptive and empirical perspective to question the purpose of the company and its policies. It studies which proportion of its results is explained by industrial policies and not by creative financial engineering, before addressing the issue of performance and results in relation to the contract. The study seeks to document these material dimensions of corporate action – investments, intervention on the network, human resources policies, and new services delivered to end users. The chapter presents two major results. First, this cooperative arrangement has improved the technical system and the level of service. Good practices circulate within and beyond the company, thus contributing improvements to the entire sector. The second result refers to the theory of regulation: the study suggests that the firm is not as free as the theory posits. Rather than a free and potentially opportunistic actor, what emerges is a constrained actor, embedded in the materiality of objects and regulations. It is thus a formatted problem that orients agendas and sets the frameworks within which the protagonists act.

In Chapter 2, Murielle Coeurdray and Franck Poupeau (with Brian O'Neill and Joan Cortinas) explore the implementation of one of the first major water socio-technical systems in the western USA, the Salt River Project, which is divided into a private water company and a public power utility. Since the beginning of the twentieth century, this hybrid entity has been challenged by several factors: funded by the Bureau of Reclamation, it aimed to satisfy the needs of local water users and specifically farmers from the region of Phoenix, in a context of arid ecosystems but also of strong and non-regulated floods. It had to respond swiftly to the concerns of communities experiencing rapid population growth, and new patterns of living have facilitated a shift from a rural use of water resources to urban use. In order to understand the adaptive capacity of the SRP model, one needs to understand how and why the technical system implemented in Arizona adapts to specific constraints at the intersection of ecology and economy. The historical perspective of this pioneering model has also to be cross-linked to a sociological analysis of specific practices of water managers geared towards combining a large set of factors – standards of technical systems; contracting and pricing issues; and the environmental constraints for sustainable growth in a context where urban sprawl has been constantly increasing – that have led SRP to defend a sustainable use of natural resources and a policy of water conservation.

Part II ("Questions of legitimacy: private interventions in a public service") analyses several examples of "remunicipalization" of water services in South America, Europe and India. In Chapter 3, Bernard de Gouvello provides a retrospective look at the concession of water and sanitation services in Buenos Aires between 1993 and 2006. In 1993, the consortium Aguas Argentinas SA led by the group Lyonnaise des Eaux (now Suez) was granted the concession of water and sanitation services for the city of Buenos Aires. This represented the most important concession ever granted to a private operator, and it was then presented as an example on the international scene. Yet, in 2006 the Argentinian authorities terminated the concession contract. The end of the concession is usually explained by the fact that the major economic crisis of 2001 put an end to the peso–dollar parity and made the economics of the concession untenable. This chapter shows that, beyond this important change in the economic context, the failure of the concession was the outcome of a deeper process of gradual delegitimation of the notion of a private operator that unfolded over the entire duration of the concession. The chapter sheds light on the economic, social and political dimensions of this process.

In Chapter 4, Dominique Lorrain studies another recent case of the remunicipalization of water services in the city of Paris where, after 150 years of partnership with the private sector, the socialist mayor decided to bring the potable water system under municipal control. This measure constituted a break with the long evolution of the "French model of urban services". The aim of this chapter is to understand what this means. First finding: the decision was not attributable to failure by the existing operators. Second finding: the change has brought no improvement in the technical system, for Parisian users and for the city government; the data on prices, investment and transparency are similar. At this point, the research runs up against certain realities – the problem of fixed costs in the water industry, inertia in public accounting categories – which do not necessarily provide a global and consolidated picture of public action. Third finding: the explanation is primarily political, the missing link in the model. The latter's constitutive principles were laid down in the late nineteenth century, and it has survived by adapting through the production of "secondary institutions" to solve problems. This process was primarily accomplished by experts – top politicians, senior state civil servants and company engineers – while users/voters were mostly left out of the decisions. Over time, the reasons for outsourcing the service were forgotten. The weakness of a doctrine that could legitimate and explain the rationality of public services contracting in France has opened up an area of uncertainty exploited by politicians. In this respect, the water sector remains a symbolic field in which divisions between left and right can still be activated. This chapter shows that even in technical activities with long time horizons, where all the rules seem to be stable, the foundational political principles need to be updated. Public action cannot be justified on economic grounds alone. It is not enough to do the job efficiently; the principle of the job must itself be legitimate, understood and accepted by users.

The case of the Compagnie des Eaux in Berlin is the object of Chapter 5, where Thomas Blanchet studies how a constrained process of partial privatization could not achieve the original outcomes of local partnership. In 1999, the city of Berlin was facing an important budget deficit and ceded 49.9 per cent of the assets of its water service company to a private consortium that included the electric grid operator RWE and the French operator Vivendi (now Veolia) for a duration of 29 years. Less than 15 years later, these two private partners have sold their assets and the company is now coming back to its municipal owner. By focusing on this episode, the research seeks to compare two models of urban services: the "French model" of public services contracting, and the German "strong local public model". It emphasizes the major role played by secondary institutions (technical norms, legal regulations and accounting categories) in framing the cooperation between these different models as well as the use water operators make of them in order to create, maintain or call into question such a cooperation. It suggests that while partial privatization has been made possible by important interventions on these institutions (legal model, rates), they have been used a few years later in order to organize the return of Berlin's water company to the public domain.

In Chapter 6, Marie-Hélène Zérah and Cécile Renouard take the case study of Nagpur's water distribution contract in order to understand and explain the effects of corporate social responsibility on public service. Indeed, since the mid-2000s, several successful public–private partnership pilot projects have contributed to establishing the subcontracting of water services as a model better adapted to the Indian context following the failures of the preceding decade. This research focuses on the first contract between the city of Nagpur and one of the big French water multinationals. It seeks to answer two questions: Can "privatized urban services" meet the challenges of sustainable and equitable development? Can they contribute to the development of local urban capitalism? The study is premised on the idea that power relations within the subcontracting company have an influence on the implementation of the contracts. The methodology thus consists in opening the corporate "black box" in order to better analyse the embeddedness of private action in relation to the state and civil society, as well as the implementation of corporate social responsibility by a multinational acting in partnership and within the framework of a public service.

Part III ("When small water systems meet large water systems") is dedicated to the different alternatives to the water regime associating a single operator with a defined territory of a "network city". In Chapter 7, Sarah Botton and Aymeric Blanc detail the role of small private operators in the supply of peripheral neighbourhoods in Africa, where the quality of public services in the cities is challenged by an increasing and non-solvable population. The extension of neighbourhoods into urban peripheries, where demographic growth is concentrated, represents an important challenge for utilities operators, since they need to develop socio-cultural access and to implement specific modes of intervention. This chapter is based on the study of three African cities (Maputo, Ouagadougou and Kisumu). It analyses pilot projects led by water operators with the support of development banks, aiming at subcontracting this public service to small, local private operators. It presents the first results of this arrangement and seeks to identify continuities as well as breaks in the commercialized supply of water in these three cities.

A different configuration is presented in Chapter 8, where Sébastien Hardy and Franck Poupeau focus on the role of cooperatives within the aggregate system of La Paz and El Alto in Bolivia to discuss the social conditions of the possibility of a self-organization of urban water management. While the development of a water supply and sanitation network in the cities of the Global South has been largely modelled after the template of European cities, with a city-wide system continuously providing significant volumes of water through a network of pipes controlled by a private operator managing the entire water cycle, the purpose of this chapter is to explore alternatives to this system and to establish whether they can be complementary. It takes as a case study the urban aggregation constituted by the cities of La Paz and El Alto in Bolivia, where many families get their water from cooperatives. The Bolivian case is particularly interesting, first because the country has been through several "water wars" that have called into question the international diffusion of the private management of resources and services, but also because it sheds light on the emergence of social practices that generate new modes of regulation, about which still little is known. In order to establish the importance of small local systems, the survey has drawn upon four sources (archives, spatial analysis through geographic information systems, questionnaire-based interviews, observations) in order to develop a geo-indexed database and to analyse the social differentiations that operate in the field of water management.

In Chapter 9, Rémi de Bercegol and Shankare Gowda study inequalities of access to drinkable water supply in four Uttar Pradesh municipalities, as a way of shedding light on the process of decentralization that has characterized late twentieth-century India and on the situation of small cities, too often neglected by researchers and public authorities alike. This level of analysis is key, since it concentrates most of the future urbanizing trends, between rural exodus and metropolitization. The chapter suggests that the subsidization of water supply is difficult to implement in small municipalities whose human, technical and financial capacities remain too weak, and whose specificities are not taken into account by water policy. In the small cities of Uttar Pradesh, the social distance between regional engineers and local plumbers, and the constraints related to the building of the technical system, contribute to explaining the difficulties of a decentralized service.

Finally, in Chapter 10, Julie Aubriot studies social mobilizations for the right to water in South Africa. At the end of apartheid, in 1994, one of the main concerns of the new ANC government was to erase inequalities and generalize access to basic services. The realization of the right to clean water, which was recognized in the new constitution, thus became a priority. The municipality of Johannesburg, which had already started redefining its water policy in the 1990s, launched the Gcin'amanzi project in its most important township, Soweto. Early on it became the subject of controversy, and in 2006 it was at the centre of a court case, the "Mazibuko case". Supported by activist organizations, five residents of Soweto criticized two aspects of the project: the installation of prepaid meters, and the free water policy, the volume of which was deemed insufficient. By shedding light on the politicization of this water-related crisis, this case shows how the reliance on law has made injustices visible, to the point of transforming significantly the social policies of access to basic services targeting marginalized urban populations. More generally, the chapter suggests that the technical and economic aspects of urban services conceal power relations in which the end user plays a prominent role.

In the Conclusion, Dominique Lorrain and Franck Poupeau develop a combinatory approach to the outcomes of the research cases presented in the book. They provide a characterization of water regimes which includes technical, economical and legal dimensions of contracts and partnerships in a broader sociological perspective on the practices of the operators and institutions involved in water supply.

Notes

1 Although the current use of the term "privatization" is used to indicate the presence of a private company, there is not much in common between a sale of assets

(land, piping, pumping and treatment equipment) and a small-scale management contract valid for a limited period of time. Relationships with the government are not comparable and the commitments of the parties involved are not the same. It would be preferable to use the three terms that cover the spectrum of these relations: "privatization" (sale of assets); "public–private partnerships", designating contracting-out (extended missions) and service contracts (precise missions); and "public management" (for state-owned and municipal companies).

2 A listed company has to deliver results if it wants to keep its shareholders. Consequently, it must compare the risks and potentials of its activity in a developing country with those of its activities in other sectors. From this point of view, the "profitability" of the water sector appears to be below the level of other infrastructure: energy, dotcoms, property, entertainment resorts. The way the private equity funds have structured their portfolio of investment in infrastructure since 2002 is an illustration of this question: not many are engaged in the water sector.

Part I

Unorthodox models and practices

1 The "embedded actor"

A water joint venture in Chongqing

Dominique Lorrain

The contracts governing the private sector's involvement in the management of networked services[1] take a wide variety of forms (World Bank, 2001; Lorrain, 2008b). In order to reduce the risks of asymmetry, of opportunistic behaviour by companies and of regulatory "capture", the economic theory of regulation has recommended a whole array of solutions: complete contracts, separation between the principal who decides and the agent who executes, independent regulatory agencies (Baldwin *et al.*, 1988). These solutions should have been enough to provide a good balance between the public authorities and private sector firms. However, they have not. One lesson learned from the past 15 years is that rates of contract failure have been high; it is not easy to navigate devaluations, changes in political majorities and economic crises. The international institutions are wondering how to find fixes for orthodox solutions that initially showed every sign of being robust (Harris, 2003; Marin, 2009).

Among the new institutional approaches, one which deserves attention is the joint venture, although it runs completely counter to the theories. The central principle on which it rests is not separation between a principal and an agent, or the signature of complete contracts. In joint ventures, the partners cooperate at all levels: they are shareholders in the operating company; they share the management functions; they operate the system in concert. The central principle is co-production. This was a solution that began to develop in China in the water sector in 2002. It was introduced after 15 years of reforms during which the Chinese authorities pragmatically explored all forms of cooperation. Within a few years this approach had increased in importance, to the benefit of French companies, but above all to the advantage of Chinese companies, new entrants to the market. It is an approach that has been implemented in parts of cities in Shanghai, Shenzhen or Chongqing, as well as in smaller cities.

Total outlier

In the 1980s, China was characterized by a model of direct administration, with little organizational autonomy, little use of price mechanisms and a major emphasis on plans and budgets (Pairault, 2008; Bergère, 2013; Sanjuan, 2006). Initially, Chinese officials turned to the industrial countries and the development institutions to obtain subsidized loans for the acquisition of technologies in all sectors. In the water sector, these loans were used for the rapid construction of potable water treatment plants to international standards. In the early 1990s, forms of partnership began to be explored – an experiment for an integrated industrial park with the support of the Singapore government, partnerships for drinking water production only – and by the end of the decade there were 25 such partnerships in the country, including nine with Lyonnaise des Eaux (Bellier, 2002: 85). In parallel, two experiments with BOT (build operate transfer) contracts for the production of potable water were conducted. One was an unofficial contract with the Thames Water company in Shanghai in 1995, which terminated after a few years, the other an official contract in Chengdu with Compagnie générale des eaux in 1998. Following these experiments and learning processes, the Chinese authorities decided in 2002 to allow partnerships extending to the whole water cycle – from pumping to distribution – in the form of joint ventures (see Appendix 1.1). In order to accelerate the introduction of urban water systems, they felt that it would be beneficial to have a single operator who would invest, operate and carry out regular modernization work. In short, they recognized the fact that the strength of a chain depends on the strength of its weakest link, and that it was pointless to finance a new plant if the water produced mixes with water from old plants, deteriorates in quality and is wasted in poorly maintained pipes.

This experiment is interesting not just because it highlights changes in China, but also for the light it casts on several theoretical currents. As regards China, the example is relevant to work on social change and the theory of path dependency (Thelen, 2003). The country had, of course, changed massively since 1992: the experiment in reform in the water sector is therefore an invitation to see what levers were employed to break with business as usual. It also shows how China has devised its own way of doing things. Initially, it mirrored international practice, but it quite quickly came to distance itself from the recommendations of donors, and to carry out reform in its state-owned enterprises (SOEs) at its own pace, despite pressure from the development institutions (Shirley, 1995). In the networked public services, it enacted a law in 1995 preventing foreign companies from having contact with end users. Reforming the SOEs to increase their efficiency would have led to massive layoffs (already substantial in the northeastern provinces) (Kernen, 1999) which, because of the lack of a social safety net, would have created risks of political instability. Similarly, introducing concession schemes at that time, like those in Latin America, would have caused numerous practical difficulties: the accounting records were incomplete, there were no regular maintenance policies, and prices remained very low. This kind of solution would therefore have disrupted the working environment for staff and would have led to reactions from users to price rises associated with the arrival of foreign firms. Faced with the choice between maintaining political balances or immediate reform, China's leaders preferred to introduce reform at their own pace, in stages, in a so-called "step-by-step approach" (Bergère, 2013). This desire for independence from the Washington Consensus is particularly apparent in the modernization of public services.

The experiment is also interesting because it challenges the common understanding of monopoly markets and their regulation. Underlying theories of regulation and contracts is the idea that the firm is powerful (asymmetry), especially in a situation of monopoly, and that it therefore poses a risk to the public authority. One way of safeguarding the public is: (1) to clearly define the terms of the problem (through studies) and to select the "right" firm (by competitive tendering); (2) to define the rights and obligations of the parties from the start (complete contract), and (3) to monitor the firm (independent regulation agency). In principle, if followed, this *doxa* is supposed to produce the right outcomes. Compared with this set of precepts which were mainly drawn up by the World Bank (Kessides, 1993, 2004; Guasch, 2004), the experiments with joint ventures in China represent a complete departure: the studies were cursory; firms were not always chosen by competitive tendering; the contracts set only very general objectives; there is no regulatory structure in the strict sense, and in fact the joint venture constitutes a structure of cooperation at all levels, in which the principle of separation between the public authority (the principal) and the firm (the agent) dissolves over time. And yet results are achieved.

Investigations in Chongqing

In relation to these dominant currents, the research we did in Chongqing[2] began with the principle that before looking at performance and outcomes under the contract, one should start with the problem to be resolved and the policies undertaken. First, what is the *history* of the joint venture and why did the local government decide to pursue that option? Of course, we looked at the formal process: the studies and selection. How were the institutions organized in terms of the contributions of capital and human resources? How were the management functions divided up? What were the terms of the contract, even if it could not be described as a complete contract? Next, in recognition that institutions provide a framework for the actors and offer incentives (North, 1990), contract theories and institutional approaches should be granted their appropriate role but no more, and our focus was on the substance of the action: the object concerned and the reality of the practices. What does the actor do? If private water firms exist, what are the specific capacities that differentiate them from any other newcomer? How much of the result is explained by industrial policies rather than by creative financial engineering? This led us to look at *industrial policies*, the big absentees in works that concentrate on the role of contracts. Are not all the factors of success to be found in these policies? While exogenous factors have an influence (major market cycles, local political factors), it remains the case that if the firm controls anything, it is these policies. That is why time must be taken to study the industrial policies pursued by the firm, rather than making the contract (what it says and doesn't say) the universal explanation for the outcomes of an action.

The substance of the survey was to document this material component of the action: investments, action on the infrastructure, personnel policies, and new services for users. This descriptive approach, which keeps at arms' length from the *doxa*, makes all the more sense in that the introduction of reform in China was not based on any fully established theoretical framework. The approach was much more pragmatic. Under the pressure of problems, relying on a few general principles, the actors explored and negotiated. What may seem to be an unorthodox solution becomes fully understandable if we consider the starting point. In other words, rather than emphasizing the "right" institutional framework that would launch the contract into orbit and provide a standard for performance assessment, and from an understanding that the approach to these problems in China is not fully predetermined, we focused on an approach via "the problem to be solved".

Chongqing is the major city in Sichuan Province, in the centre of the country, at the intersection of the Yangtzi (Changjiang for the Chinese) and Jialing rivers; the terrain is steep, the river wild at flood periods. In May 1997, the central government gave it the status of a national central city, alongside Beijing, Tianjin and Shanghai. This "greater" Chongqing extends over an area of 82,300 km^2 and at this time had a population of around 30 million, distributed across several cities (Chonqing, Fuling, Wanxian, Qianjian) and rural districts. The city of Chongqing – if we define a city as a continuous built-up area – then had three million inhabitants, but this figure rises to more than five million if we add together its urban districts (which include rural areas) and if 15 million inhabitants if its rural districts are included. In 2010, the population of greater Chongqing was around 34 million, and that of "urban" Chongqing five million. All in all, the municipality controls a very large area, with a population of around half that of France. The city can grow through demographic growth, but also by incorporating territories previously classified as rural areas.

The institutional upgrade to national central city status in 1997 took place against the background of the Three Gorges Dam Project. The central government wanted to establish a framework that would give it direct control, without going through the province. Apart from the building of this massive structure, the aim was to run a large number of supporting projects. As the water level rose, several cities would be flooded and rebuilt. To prevent outflows from the catchment areas turning the reservoir into a gigantic pollution trap, public investment programmes were needed, with the additional goal of developing the centre of China in order to keep part of the rural population in place. From this time, therefore, it was known that Chongqing municipal authority would need to invest considerable sums in its infrastructures, and in particular in the environment. The World Bank conducted studies and granted loans for several projects. In Chongqing, it supported a massive anti-pollution plan, consisting of an initial phase in which interceptors would be built, followed subsequently by the construction of a wastewater treatment plant. While the potential problems were diagnosed fairly quickly in the years 1997 to 2000, implementation had yet to begin. At this time, Chongqing municipal government was still operating through direct administration. Its organization was hierarchical. Most of the municipal departments received their resources from the finance bureau or from other public transfers. As a public system, the accounting framework included no depreciation or autonomous operational units (such as cost and revenue centres). To obtain revenues, the authorities created a range of additional taxes, but nobody knew whether they were "fair" for the inhabitants or how much they cost to collect.

Within a few years, Chongqing underwent considerable changes, both physical and institutional. The city grew and attracted investment. The central part (between the two rivers) bristled with skyscrapers; the urban space was criss-crossed with bridges and elevated urban expressways; a light subway was built; and development also took place along the north bank of the Jialing. The old Jiangbei district was renovated, taking advantage of the deconcentration of state services which were moved out of the old centre of Chongqing. Further north, business was boosted by a production joint venture with the Ford Corporation. In other words, within a few years, the capital of Sichuan emerged from its torpor. At institutional level, the municipality undertook reforms: its public service boards were converted into companies; it set up development and investment companies (Chengtou) through which it could raise capital in the form of bond or share issues. The 1993 accounting reform was applied. This was the background to the establishment of a joint venture with the Sino French group for drinking water operations (see Appendix 1.2).

Setting up the joint venture

In November 2002, Chongqing municipality formed a joint venture between its water company (Chongqing Water Group)[3] and Sino French: Chongqing Sino French Water Supply Company (see Appendices 1.2, 1.3 and 1.4). It granted this new company an exclusive 50-year right to operate the potable water service for Jiangbei, an urban district located north of the historical centre, on the other side of the Jialing River. The contract also included the plan for very early expansion to encompass a town in the Yubei district near the airport, and the development zone where the Ford Corporation's automobile production joint venture is located. This operation, which received less media coverage than the Shanghai Pudong contract, illustrates the readiness of the Chinese authorities to explore the partnership option as a way of running an entire urban service.

There are several explanations for the contract. The municipality was growing so fast that it had to modernize its urban services quickly. This partnership was a way to mobilize financial resources and acquire operational expertise without losing control: the joint venture maintained a balance of power between the two partners. Moreover, it only applied to part of the urban area. Most of the city continued to be served by the municipal board (Chongqing Water Supply Company), a subsidiary of its public holding company. The contract for the northern part would act as a catalyst for modernization. The operation may also be considered from the perspective of Sino French and the development of Suez Environnement (originally Lyonnaise des Eaux) in China. The company had been involved in partnerships for the production of potable water in the country since the early 1990s and was looking for an opportunity to demonstrate its expertise as a service operator, by raising quality while maintaining financial equilibrium.

> We bring few major technologies, but we are better than a public company at managing a large technical system, in which quality is made up of lots of little things. Part of the experiment is to introduce day-to-day operating skills.[4]

In addition, the city of Chongqing was not entirely new to the company. In 1999, Degrémont (an affiliate of Suez) had many reference points in China – the design and build of 93 water treatment plants (Degrémont Beijing, 1999) – and had constructed a small 48,000 m^3 unit in Chongqing as part of the 1980s master plan prior to the Three Gorges Dam project. Finally, there was also one personal factor: a new deputy mayor had been appointed in the city a few years earlier. Originating from Shanghai, he was convinced of the benefits of involving the private sector and had a positive view of Suez and its CEO Gérard Mestrallet. For all of these reasons, the municipality and Sino French began direct multi-criteria negotiations at the start of 2002, completed in November.

With regard to the capital structure, the joint venture was 60 per cent owned by Sino French Water Development, itself owned in equal parts by Suez Environnement and New World. The 40 per cent Chinese holding was owned by Chongqing Water Group (CWG), whose own share structure was changing. At the time of the agreement, CWG was 100 per cent controlled in traditional fashion by the municipality's assets committee (Sasac). In April 2008, Suez Environnement and Sino French acquired 15 per cent of its capital, an operation that reflected the consolidation of the partnership. On 29 March 2010, 10 per cent of CWG's capital was floated on the Shanghai Stock Exchange, in a successful operation that ended in a new shared ownership structure – 76.5 per cent Sasac, 10 per cent public, and 13.5 per cent Suez Environnement and Sino French (see Figure 1.1).

The Chinese side contributed to the capital by bringing existing assets[5] and the private partnership with HK$800 million, subsequently reinforced with a further HK$200 million for new equipment. For valuation, the assets were audited by the Sino-French teams and by a Chinese company. Thirty-five per cent of upfront expenses were covered by Sino-French (HK$280 million), a figure that corresponded to its share of the company's net assets,[6] and each

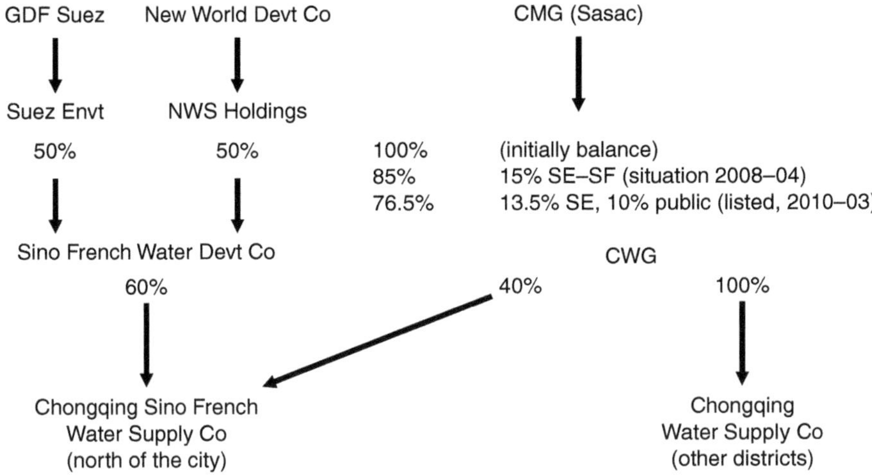

Figure 1.1 Control of the two Chongqing water companies.

of the two shareholders contributed half. The remaining 65 per cent were funded with loans from Chinese banks, to be repaid out of operating profits. In 2002, financing on the national market was something new for this kind of project, since finance had previously been raised via the international markets, i.e. in dollars. There was nothing theoretical about the resulting currency risk, as several experiences show: Argentina, Mexico, Philippines.[7] The Chinese banks were interested, because they were "*looking for sound, long-term projects*".[8] Sino-French was both a shareholder and operator of the water company. An interim agreement between the stockholders set out the distribution of responsibilities. The board of directors consisted of eight members (three on the Chinese side, five for Sino-French). The chairman represented CWG. The CEO of the company was appointed by Sino-French, subject to acceptance by the Chinese side, which also chose a deputy CEO and CFO (chief financial officer). The president of the (very influential) employees' union was appointed by the CCP (Chinese Communist Party). The other directors were chosen by the CEO, on the basis of consensus: "most important is a good cooperation between the partners".[9]

The agreement was an incomplete contract:

> There are not too many guidelines set up in advance and which are compulsory. The JV has some broad (general) obligations to respect: supply enough water with a good quality. But in the same time it is clear that the company is not totally free.[10]

As in the case of a lease, the firm is managing the whole technical system and has to make the decisions necessary to meet the population's needs and to provide the service. In a context of rapid growth, as was the case in Chongqing, this obligation is anything but empty words. The company's behaviour is largely determined by the situation in which it operates; the primary principle stated in the contract – "to satisfy needs" – represents a powerful steering mechanism, limiting the firm's freedom and thereby the possibility of opportunistic behaviour associated with such freedom. Over the first seven years, fulfilling demand was enough of a goal to employ all the company's energies: "we bought two plants (in 2003), we invested rapidly in a new one in 2006 (160,000 m^3/d) and another is under construction (200,000 m^3/d)".[11] In February 2003, as planned, the contract was extended to the Yubei district, where the infrastructures required upgrading. Located 20 km north of the centre, this was a very fast developing district, one of the development hubs specified in the master plan.

Industrial policies

The process was fairly typical, consisting of three components: modernization of the technical system, relations with users, and operational reorganization. These were the policies needed for the company to meet the terms of the contract, establish its legitimacy and (in principle) make a profit. Contrary to analyses of monopoly markets, profit here is not the result of opportunistic behaviour during operations, or the exploitation of an asymmetry in capacities at the time of the signature of the contract. It should rather be interpreted as the ex-post culmination of industrial policies. The annual statistics for production and sales, for the length of the network and for the number of water meters reflect the conditions of growth (see Table 1.1). Such a situation both provides an answer to the problem of regulation – the controlling factor is the problem to be solved – and also contributes to the financial equilibrium of the contract, because in this type of industry a rapid rise in consumption is the quickest way to absorb fixed costs.

The first of the policies was to keep pace with growth and *to make investments*. Production doubled, with average annual growth of 12 per cent (see

Table 1.1 Basic data

	2003	*2004*	*2005*	*2006*	*2007*	*2008*	*2009*	*Multiple*
Production[a]	190	223	252	272	322	349	380	
Water sales	145	167	188	221	245	275	308	2
Pipes > 100 mm[b]	272	392	542	677	768	877	940	3.5
Meters[c]	36.8	92.1	153.3	199.0	261.2	319.0	369.0	10

Source: company, data compiled by author.

Notes

a Production and sales in thousands of cubic meters per day.

b Length of pipes in kilometres.

c Number of meters in 1000s.

Table 1.1). The length of the distribution network (pipes with a diameter of over 100 mm) was increased by a factor of 3.5. The number of water meters grew tenfold. This exponential growth is partly attributable to new regulations enacted in the first year of operations, the so-called "one home, one meter policy". Whereas the volume of water consumed in apartment buildings had previously been measured by a collective meter, the new policy promoted individual meters and made it compulsory that they should be changed every five years, and every three to four years in the case of the biggest meters. In truth, users were given the choice between keeping the old collective meter or changing it for a new individual device, at a cost of 900 Yuan. According to our interviews, most users chose to replace on the grounds that it was to their advantage to be billed for their real consumption, since some building managers added charges to the bill issued by the company. To facilitate the transition, the municipality supported this policy, giving poor households subsidies of up to 500 Yuan (€55). This shift to individual meters is also an interesting example of the notions of incompleteness and asymmetry. Incompleteness in the contract, since the new rule was introduced in 2003, at a time when the two parties had just completed more than nine months of negotiations. Asymmetry, but not in the way expected. In reality, the firm had no choice but to adapt to a unilateral decision by the public authority. The application of this rule generated costs that were not included in the contract: costs for installing meters, sometimes for changing pipes, and an extension of the company's responsibility for the cost of leaks.[12] At the same time, this policy was a powerful factor in developing knowledge of the network and driving action to reduce the rate of losses: a forced incentive, one might say.

In 2007, the loss rate was 23 per cent.[13] Measures to maintain the system and replace pipes and meters reduced wastage to around 20 per cent in 2010. However, according to the operations director, this level was still high compared with other Sino-French joint ventures. The company set up a team of inspectors to track consumption and check abnormal fluctuations. The big meters were read every month and the others every two months. This effort in preventing water losses was strategic. It was a matter of image; improving efficiency is an environmental protection measure. In addition, reducing the volume of water withdrawn, treated, transported and distributed cuts operating costs. Finally, this ratio is one of the criteria taken into account by the municipal government when the company submits a request for a price increase; if the ratio were too high, the municipality would require efforts to reduce costs, before granting an increase in revenues.

The establishment of a laboratory in order to monitor *water quality* was one investment that took place in the first year of operations. The joint venture received support from the group's research centre. This lab tracks 90 parameters, 79 of which are among the 106 standards comprising China's new regulations. For the other parameters, the tests are done by Chongqing Water Group's central laboratory. On average, the compliance rate is 99.9 per cent.

In 2010, there were 59 sampling points (50 on the network, four for raw water entering the system and five for treated water). This procedure is followed 365 days a year. In addition, two municipal government offices carry out their own tests: the joint venture is given no warning of inspections, and the sampling points are different.

Very early on, in September 2003, the joint venture set up a *consumer service centre*. This centre is open every day and provides a full one-stop service: opening/closing accounts, payment, bill explanations, leaks, meter replacements and so on. Users no longer need to go to different offices. A call centre was also opened, together with a 24/7 hotline. Finally, as with other group operations, a satisfaction survey is conducted by an independent body. Unsurprisingly, for the first nine years of the contract, the satisfaction curve followed a trend already found elsewhere: very sharp growth at the beginning, then stability or even a slight downturn, a tendency summed up philosophically by one of the people we spoke to: "The more you improve, the higher are the expectations".

One very important measure related to *company reorganization*, with the shift from a hierarchical structure to a multi-divisional form, itself part of a local–global structure. This was a way to underline the new priorities: placing maintenance and user service functions on a par with production (and its technical approach). This reorganization also made it possible to promote certain members of staff, a difficult task in a personnel management system that is very much governed by formal and informal rules. Previously, each potable water production plant had formed a mini-company with its departments: production, pipe laying, operations, finance, management, administration. This organization comprised four levels: plant director, office manager (*office/Ke*), shift manager and worker. The big organizational change aimed to eliminate the plant-based structure and to organize the company around eight departments,[14] the largest being subdivided into offices. The position of plant director disappeared. The new structure has five hierarchical levels. The creation of department directors, below the CEO, produced a management structure that offered prospects for promotion. This structure is a way to concentrate skills around certain objectives: effectiveness in expanding the network, finding leaks and maintenance, reading meters, and service to users.[15] It also makes it easier to implement a policy for the whole service delivery area. In 2003, following the takeover of two plants, the company had a workforce of 530, producing 190 million cubic metres over a supply area of 75 km^2. By 2009, production had doubled, the distribution area had grown to 104 km^2, but the workforce remained at 550.[16] The joint venture had made productivity gains.

One component of the reforms concerned the *personnel*, since the creation of a private company – and a foreign company at that – was a big change for employees.[17] At the same time, under the existing system, they had the legal means to fight their corner, to block an operation or at least to make the management of a company very difficult. Workers had their own reporting system through which grievances could be passed up through the levels right to the CCP's human resources department. These staff members were civil servants, with a status, legal protections and a pension system. For this reason, reaching a consensus was a crucial factor in the management of the company. At the beginning, many worried about their future. They were concerned about their status and the risks of redundancy associated with an international employer. In fact, these fears were to prove groundless. Employees of the public company with more than 15 years of service could retire under the public system; the others would have to move into the private system or resign. This transfer period went smoothly. The Chinese managers whom we met concurred: the company's management cooperated with the other stakeholders.[18] With regard to wages, there was very little change for equivalent work, since any increase had to be approved by the elected officials and applied to both water companies. This made for a level playing field. The board of directors approved the overall wage bill. After this, the distribution was down to management, but since it had to apply the rules, it had limited leeway. Individual improvements resulted essentially from the creation of new posts. Previously, wage differences between employees had been very low: between ten and 20 Yuan a month. With the new structure and bonuses, the differences could become wider. An assistant director might receive an increase of 800 Yuan per month above his previous position, for an office manager the increase could be 150 Yuan per month, and for a shift manager 50 Yuan per month.

Personnel policy also applied to *training* and to a *workplace safety* programme. Once again, we see the importance of the rules, or structured problems, as frameworks, and their importance in preventing opportunism. Every year, the department directors decide a training plan with the people concerned, the training goals and so on. This plan is sent to the company's assistant director, who makes the decision. After training, employees complete an evaluation on the quality of the programme, and sum up what they have learned. The number of people trained is increasing, according to the president of the union, and staff are satisfied, seeing this as a benefit they receive from the joint venture. The nature and locations of training can vary considerably. For example, in 2009, "all team managers received one week's training at Chongqing Water Group headquarters".[19] Executives or technicians may go to Macao or to another Sino-French operation.

Health and safety policy was markedly developed. Existing rules were reinforced. A health and safety programme was implemented in 2006, a worldwide programme applied in all of Suez's operations and one of its non-financial undertakings. What we see here is an illustration of the emergence of the theme of corporate social responsibility alongside financial reporting (Lyonnaise des Eaux, 2008; Suez Environnement, 2010a, 2010b). The process may be broken down into several components: creation of a health and safety committee, establishment of safety routines, with a total of some 15 criteria to be met. The company then audits the programme. The CEO is required to go on site once a month to observe whether or not the rules have been followed.[20] These requirements now apply to subcontractors, which obliges the joint venture to be more vigilant: "we make a risk assessment; we sign a contract with the subcontractor regarding what he has to do concerning safety".[21] At the time of our first visit, there was a big display panel stating: *877 days without injury*, a figure which, at our second visit, had risen to 910 days. Seen by all employees when they enter the main site, this sign means that the company has not experienced any injury caused by a workplace accident whether in the plant, in a pipe-laying operation, in the lab or during business travel. Initially, employees were sceptical about the benefits of following the rules, but over time they became convinced that it was "to protect them".[22] This global programme is approved by the board of directors, and its subsequent implementation is the responsibility of the CCP union. Chongqing Water Group decided to introduce this programme into all its subsidiaries, and the joint venture is its model unit. In November 2009, the company received ISO 9001 certification for its management procedures.

The joint venture in practice

The joint venture's 2008 revenues for water sales were around 200 million Yuan, added to which around 50 million Yuan were for maintenance and construction works. In 2010, the figures were about 220 MY and 80 MY, respectively, increasing revenues to 300 MY. The dividend paid to shareholders (i.e. the rate of return) is set by national rules as a percentage of the company's net assets, varying between 8 per cent for a Chinese company and 10 to 12 per cent for a foreign company. Previously, the company had made a loss.[23] With net assets of around 450 MY, the potential dividends that the shareholders (CWG 40 per cent and Sino-French 60 per cent) were allowed to receive, stood respectively at around 15 and 30 million Yuan per year. The officials whom we met did not give us the precise data on this point (nor did the other surveys of joint ventures), but we established that the company had paid dividends every year and that the amount had varied markedly from one year to the next. "The directors are reasonable in their dividend policy, everything depends on the investments that the company needs to make".[24] Decisions are based on results and the investment programme. Every year, the company office approves an investment plan which can be partially funded from borrowing, provided that the debt does not exceed 65 per cent of assets (after depreciation). This rule sets the maximum that the company can borrow and therefore the level of dividend reinvestment that is desirable or necessary. Once again, therefore, what we find is an actor that is firmly guided by the rules; it does not operate in an unrestricted space that could theoretically lead it to maximize its profits and strip the company of its earnings. Quite the contrary, it operates in a space of constraint. Its starting point is one primary goal – to develop the technical

Table 1.2 The price of water in Chongqing (yuan/m³)

	2002	*2010*
Eau	2.1	2.5
Sewerage	0.7	1.0
Total	2.8	3.5

Note
Price for domestic consumers accounting for 58 to 59 per cent of total sales.

system – for which the necessary investment is set aside; then the required debt ratio sets the volume of maintenance and construction work financed by borrowing and by shareholder contributions. Dividend distribution is the last of these priorities.

Between the start of operations (November 2002) and 1 January 2010 – when the tariff increased for the whole municipality – there was no change in price. This increase followed the standard procedure – an initial proposal by Chongqing Water Group, then studies by the price bureau followed by a public hearing. This hearing was held in November 2009 and the 26 people present were all representatives of institutions. By contrast with other cities, no lottery was held to select ordinary citizens to attend.[25] There was only one vote against the increase. The additional revenues were immediately allocated to investment. A programme to replace old pipes was drawn up with Chongqing Water Group (CWG), with the target of replacing 80 km of pipes over a number of years. The joint venture maintained operational control, whereas the public holding company took charge of planning, coordination and interface with the municipal offices (see Table 1.2).

Ultimately, despite no price increase for seven years, the company managed to invest in the network, train its employees, improve the service and pay dividends to its shareholders. This outcome deserves attention, because it casts light on the mechanisms of performance and helps us understand the underlying roots of how the major firm works. Ignoring industrial policies, contract and natural monopoly theories concentrate on outcomes and distribution; in other words, on profits.[26] This simplification opens the way to the exploration of ideas of cooperation inspired by game theory. However, this representation of phenomena only partially accounts for the real world, because it reverses the order of things. Before we talk about the allocation of profits, we need to understand how they are achieved, and why, in similar environments, some organizations will shine and others fail. The example of this joint venture demonstrates that there is not just one miracle formula, but that outcomes are achieved by gradually improving a number of factors. A description of the different industrial policies pursued by the joint venture allows us to identify all these micro-levers. None is more important than all the others, because they are all linked: trust between the shareholders, commitment by management and the entire workforce (which comes through real policies, namely training, reorganization, salaries, health and safety), the firm's professionalism and its understanding of the right way to manage a network. Compared with other sectors, this activity is "low-tech", yet all the different micro-skills employed by the firm hold together and explain its achievements.

Back to theories

In order to understand the logic of research on monopoly markets and contracts, we need to go back to the way economists conceive of action. This concept, when applied to public–private partnership situations where the asymmetries are substantial, naturally leads to the quest for safeguards. These recommendations, translated into practical arrangements – i.e. second-rank institutions (Lorrain, 2008a) – are then spread around the world by development institutions and the finance and consultancy industry. This example of a joint venture in China – given the size of the country and the very heterodox character of the partnership – forces us to think about the sequence of ideas that takes us from the broad picture of economic exchanges to situated instruments.

Concepts of action and of actor behaviour

The standard economics model is of a flat world, without rough edges or history, in which two actors negotiate a transaction. They are rational, have access to information and seek to maximize their gain.[27] The influence of this model has been considerable, probably because its formal expression makes it easy to grasp. It puts forward the idea of a strategic actor with absolute freedom of action, which can decide to act for the common good or for its own advantage. This raises the spectre of opportunism, further amplified when the asymmetry is large and particularly in the case of monopoly. If we accept the view of action under this standard model, therefore, the logical sequence of freedom–strategy–opportunity necessarily prompts the development of safeguards to protect the weaker actor, and hence the public interest. Indeed, this vision of a flat world with instant trading and the risk of opportunism is a fair reflection of the modern finance industry, with its global market, 24/7 transactions and order-and-click trading systems. However – and here lies the paradox – the finance industry escaped the vigilance of the economists of regulation, gaining acceptance for its vision of market and professional self-regulation. As Joseph Stiglitz pointed out when Congress was thinking about setting rules, the response was invariably "trust us" (Stiglitz, 2003: 11).

Conversely, the regulations designed to counteract the risks of opportunism were essentially applied to the world of networked industries and utilities, sectors that carry a heavy load of constraints by the very nature of their activity: (1) large-scale and long-term fixed assets; (2) prolonged presence in a territory; (3) return on investment that comes over the term of the contract; (4) the threat of a public partner which can unilaterally change the rules, and (5) a proliferation of second-rank institutions that frame and manage events. This study of the real practices of the actors in this water management joint venture demonstrates that real-world principles are not entirely those of the theoretical construct, on three levels.

First, the world has a history – there are always rules for organizing exchanges, even in situations where no property rights exist (Ingold, 2008). This joint venture shows how many rules there are: health and safety standards, personnel management rules, urban planning and maintenance and construction rules, price review procedures and so on. These rules exist outside the framework of the joint venture: they would apply to any operator, even one that was legally independent. Second, the actor does not operate in situations of total freedom that would allow it to choose among different options and maximize its profits. The joint venture entails rules that define a space of action with constraints: the requirement to support the development of the fast-growing city while maintaining a particular debt ratio. The idea of the free and potentially opportunistic agent contrasts with that of the restricted agent weighed down by the material ballast of things and rules. This agent has to tackle problems and knows that it will be judged on its ability to do so. The imperative of the problem is the dominant factor, the factor that steers the agenda. And third, the actors are not necessarily governed by the urge to maximize profits.[28] Every party has its own objectives: some large firms have long-term strategies and worry about their reputation; some politicians work for the public interest and are sensitive to their historical legacy. All these factors combine and significantly limit deviance, even in the case of powerful monopolies.

The principal–agent theory and regulation

The agency theory at the heart of regulation theories (Jensen and Meckling, 1976; Laffont and Martimort, 2002; Stiglitz, 2001) works on a principle of separation and asymmetry. In partnerships between the public authority and the firm, the assumption is that it is the public authority which is at risk. What we see in this joint venture is somewhat different. First, asymmetry does not necessarily work solely to the benefit of the firm. True, the latter dominates in terms of size and the resources it can mobilize at group level, but the public authority is not bereft of resources either. It has political resources that enable it to propose or impose its views. It is also clear that the partnership functions, in this case as a mechanism for the transference of skills from the JV to the public company operating in nine of the 11 urban districts. The partnership with a foreign actor is mirrored by a concomitant public structure which captures and reproduces innovations.[29] Thus the asymmetry is reduced. Moreover, once it has acquired a certain level of expertise, the public side can stand on its own feet if it wishes. The firm knows this, and behaves accordingly. The research also shows how the firm has to accommodate to its environment. This joint venture is embedded in two worlds: the global corporate world and the world of local institutions. Its membership of the former is manifested in a few "corporate" programmes – support from experts and certain reporting requirements (Lorrain, 2007) – but none of its policies reflects a corporate protocol. Its action is always adjusted to local conditions. For this reason, membership of the "local" world is expressed in the power of rules that sharply mitigate the power that may naturally be attributed to the firm. Cash flow is managed locally, together with loan negotiation and the choice of suppliers; there is no corporate policy, even at the level of Sino-French, which is the holding company for activities in China.

The theory sees the firm as a foreign body in relation to the public authority, capable of conducting its own strategy. Here, however, the firm is deeply embedded in a local field of forces. Information and decision-making are shared within the board of directors, and it should be added that – in the event of major conflict – it is hard to imagine a foreign firm suing a Chinese political institution. Powers are shared at all levels of the joint venture. There are permanent exchanges with Chongqing Water Group in every area (quality control, staff training, joint management of a works programme, health and safety). Very precise rules have to be followed at each stage (price rises, wage policy, investment planning). The reality of action in China is one of a multiplicity of subtle rules, applied by agents familiar with every detail (formal and informal) and capable of using this power to facilitate or control events. Thus the private sector actor has to make accommodations at all levels, from the board of directors down to obtaining the most basic certificate. This reality corresponds to an image that is the reverse of the blank slate picture of supply and demand, in which the strongest player is in a position of absolute freedom.[30]

Ultimately, this joint-venture experiment shows how behaviours differ in several respects from those suggested. First of all, the public authority holds a number of cards. It can unilaterally change the rules of the game. Conversely, the large utility firm is not all-powerful and cannot behave opportunistically, because there is nothing nomadic in its industrial model. It is in place for the long haul, and will only obtain a return on its capital over time. In the event of a conflict with the public authorities, it is therefore more vulnerable than the theory presupposes. Next, with regard to the possibility of establishing a complete contract from the start, this experience is not based on prior assumptions, but takes the form of a process. The people whom we met show that they did not set off with preconceived ideas. They started with problems that they needed to solve and wanted to build a relationship that would lead to a transfer of skills. This idea is important, because in the one case there is an assumption of existing knowledge – which can be bought and distributed as a product – whereas in the other case the action is conceived of as a shared pathway that is in part a process of discovery, so not everything is calculated in advance and mechanisms of exchange are crucial in the course of the action (the reverse of the separation between principal and agent).

Finally, with regard to regulation, the joint venture is a system regulated by means other than the combination of a complete contract and a regulatory agency. Regulation in this system is partly "top-down", based on general rules (satisfying people, obeying the laws for public services), and partly "bottom-up", based on problems to be solved which are themselves highly structured by rules (product standards, principles of corporate management, employment rights, price- and wage-setting procedures). Operating within this interwoven framework of public regulation and problem-solving, the firm cannot do just anything it wants; in fact the firm is embedded or ballasted (to use a maritime image). This reality markedly qualifies the argument about the risk of opportunism. It differs from the complete contract approach in terms of the time and manner governing the use of the rules. In the orthodox view, this relation is calculated in advance: it is assumed that everything can be anticipated and managed by guidelines for the parties involved. In the heterodox approach, the actors commit to global goals, know that they can rely on a number of specific rules to decide details and negotiate conflict of interets and, for the rest, agree to manage their relationship in terms of the nature of the problem to be solved.

Appendix 1.1

Standard mounting of a joint venture

A municipality converts a public company into a corporation and sells part of the capital to a private partner. The balance may vary from 50/50 to 20/80 depending on the quality of the assets, the tasks assigned to the firm and the balance of power. The municipality pays for its equity by contributing existing assets. For its part, the private partner contributes cash based on the value of the assets, possibly including an intangible value based on the potential that the company believes the operation to have. The joint venture's board functions are shared. The private partner is responsible for management.

Appendix 1.2

Sino-French

Sino-French is a subsidiary of Suez Environnement (originally Lyonnaise des Eaux) and of New World (a major Hong Kong developer). This equal partnership was established in 1985 to take over the Macao water company and manage the water service for the city, then under Portuguese control. This partnership endured, and generated a lot of results. Twenty-five years after its formation, Sino-French has revenues of around €720 million, divided between Sino-French Water Development (€400 million), Macao Electricity (€276 million) and Macao Water (€40 million). In China, the company operates 24 contracts in 18 cities. All these contracts are joint ventures between a municipality and Sino-French.

Source: Company, May 2010, interview with the CEO of Sino-French.
Didier Rétali, De Macao à la Chine, In D. Lorrain (dir.), *Gestions urbaines de l'eau*, Paris, Economica, 1995: 205–212.
Oscar Chu, Les eaux de Macao en courbes (1985–2001), *Flux* no. 52/53, April–September 2003, pp. 117–120.

Appendix 1.3

Chongqing Sino-French Water Supply Company in 2010

Location of operation: Jiangbei district, Lianglu city (Yubei district), a development zone.
Population served: 1.2 million inhabitants (460,000 in November 2002).
Area served by networks: 104 km^2, total area 400 km^2 (area served in 2002, 75 km^2).
JV between Chongqing municipality (40%) and Sino-French (60%).
Contract set up November 2002, duration 50 years.
Water obtained from Jialing Jiang river.
Loss rate (NRW) 20%.

Appendix 1.4 The technical system

Technical system	*Production capacity (m^3/d)*
– At time of contract signature, 2002	260,000
Plant 1 and 2 on the main site (Liang Tuo)	100,000 + 100,000
Plant 3, other site	60,000
– Expansion of range and takeovers in 2003	380,000
Plant near airport 70,000 m^3/d and industrial zone	50,000
– New plant in 2006 on the main site, 160,000 m^3/d	540,000
– New station, 200,000 m^3/d, delivered in July 2011 built by the investment JV in Yuelai, upstream on the Jialingjiang	740,000

Notes

1 These include infrastructures, public services, utilities and universal services.
2 In April and May 2010, I conducted research in four joint ventures in China, two in which the partner was Towngas and two others Suez Environnement. This operation was funded by the Research Department of AFD (French Development Agency). Chongqing was chosen because I had already done research in that city in 1997 and 1998 and wrote a report on behalf of Sogreah on the water and sewerage institutions, studies that were part of the World Bank's support programmes

for the Three Gorges Dam. I would like to thank the following for their support and their comments on an initial report: Aymeric Blanc, Nils Devernois and Véronique Sauvat (AFD), the teams from Suez Environnement (Charles Chaumin, Frédéric Gourdin) and from Sino-French (Steve Clark, Zhu Shu Cai, Fu Seng Lek, Serena Qian), and all those who agreed to talk to us. Finally, my thanks go to my colleague in Tongji, Dr Yang Menying.

3 In 2010, Chongqing Water Group (CWG) was one of the eight municipal companies. Its original name was Chongqing Water Holding Group.

4 Interview with the first director of the joint venture, 2004.

5 Three water treatment plants, the pipe networks and the municipality's technical system, added in February 2003 (see Appendix 1.4).

6 The net assets were some 450 million Yuan, i.e. 270 million Yuan for a 60 per cent holding.

7 In 2002, Suez Environnement found itself facing a crisis in Argentina following a devaluation that had reduced the value of the peso from 1 to 3.5 pesos per $1, with a corresponding impact on the debt level of its contracts. The importance of financing by national credits to the long-term balance of the contract is clear.

8 Interview with the first director of the company (summer 2004) and the current director, May 2010. See the financial press on the situation of Chinese banks and the problem of "bad" debts. The large banks were floated on the stock exchange: China Construction Bank, Bank of Communications in 2005; ICBC, Bank of China in 2006.

9 Interview, CEO, April 2006.

10 Interview, CEO, May 2010.

11 Interview, CEO, May 2010.

12 Since the cost of leaks was allocated from the meter, the cost of leaks occurring between the old meter (where the network entered the building) and the new meter (each apartment) was now borne by the company.

13 Comparisons with previous years are not very relevant, because the perimeter of intervention was changed by the incorporation of business zones and former rural networks, which might be expected to increase the loss rate. During our research, we encountered the same problem in Wujiang, a town in Suzhou municipality (Jiangsu), where the loss rate in certain rural districts could be as high as 70 per cent.

14 Interviews, April and May 2010, and the joint venture's website.

15 Interview with the former director of the 60,000 m^3/day plant, subsequently director of the consumer department.

16 This figure is itself the outcome of retirements (80) and recruitments (100), the latter including graduates and former army members who had passed a test and received training.

17 Interviews: assistant general manager and secretary of the company's CCP, president of the union, chief executive officer, April 2010.

18 President of the employees' union, 7 April 2010.

19 Ibid.

20 The wearing of safety helmets and clothing, and eye protection for lab workers.

21 CEO, 6 April 2010. Note that in the international debate on CSR, the inclusion of outside contractors is one of the criteria used to assess the genuineness of the policies and to differentiate between hype and real changes.

22 President of the employees' union, op. cit.

23 Secretary of the CCP for the joint venture, 6 April 2010.

24 CEO, 10 May 2010.

25 Eleven people representing consumers from the 11 districts (nine for CWG and two for the JV), three people representing the operators, ten the CCP and the municipal departments, and two experts.

26 It is perhaps no accident that in day-to-day management language people talk about the importance of the "bottom line".

27 For a critique of this model applied to the electricity sector in Great Britain see Defeuilley (2009).

28 See the work by Herbert A. Simon, which won him a Nobel Prize in 1978, and his distinction between maximizing and satisficing.

29 In Chongqing, the authorities are reproducing the scheme already introduced in Suzhou in the mid-1990s.

30 For an interpretation that shows what behaviour can result from the reverse situation, namely a complete absence of rules, see the history of the first British water company from 1619 to 1904 (Defeuilley, 2013).

2 "Delivering more than water"

The Salt River Project: the invention of an adaptive partnership of water management

Murielle Coeurdray, Franck Poupeau, Brian O'Neill and Joan Cortinas Muñoz

> SRP founders were committed to supporting economic development, serving as stewards of natural resources and improving the quality of life in the Valley. The company continues to act from this same powerful vision today.
>
> (Letter by the President and Vice-President of SRP (SRP, 2014a))

One would think that defining whether a water distribution system is a public or private operation could be determined from the status of the operator or from the contract between the operating company and its authority. Although neither approach is incorrect, neither is always sufficient to determine the private or public character of the practices. The private–public dichotomy seems particularly inappropriate for describing and understanding technical networks where user associations play a leading role as they combine characteristics of each category: they hold shares of an operating company (apparently private) that is not profit oriented (apparently public). This kind of organizational model has been implemented by small cooperatives mostly in emerging countries to provide an alternative to incumbent bureaucracies or large private companies. These cases have attracted most of the attention and are quite well documented[1] (World Bank, 2006). But, this cooperative model is also seen to have come into existence for the development of large technical water systems in the USA and particularly for the mega-infrastructures (e.g. dams and canals) that emerged in the West for over a century (Gottlieb and Fitzsimmons, 1991; Pincetl, 2011). This may be surprising, since this part of the US traditionally evokes the ideal of the frontier, a place where pioneers were settling Western lands, leaving the crowded and industrial East for the conquest of wild spaces (Cronon, 1991; Garreau, 1991; Duany *et al.*, 2001). But the West has still been the place where water mega-projects were promoted and supported by the federal and state governments to help local communities and economic organizations operate technical systems to cope with an increasing agribusiness and sprawling urban growth (Worster, 1986).

This cooperation among different stakeholders – government administrations, utilities, irrigation districts and user associations – is presented as a model of management of natural resources, which deserves to be examined because it dates back to the early 1900s and has inspired the implementation of existing delivery systems of water and power, as illustrated by the case of the Salt River Project (SRP) in Arizona. SRP is the name used to refer to two entities that provide water and electricity to Phoenix, one of the most sprawling urban centres in the western USA (see Box 2.1). Created in 1903 as a land-based organization, the SRP model generally raises questions about its capacity to meet the needs of urbanizing communities (Logan, 2006). The way in which this model responded to the concern of communities experiencing rapid population growth and new patterns of living has facilitated a shift from a rural use of water resources to urban uses. This adaptation issue has to be understood in light of a broader historical context of water use. Indeed, water managers in Arizona dwell on the fact that their know-how in facing water scarcity has not come out of the blue but rather is based on a broad-ranging experience gained over the past century, which is the very basis of their professional legitimacy – a collective accumulation of practical skills and science based on decision-making.

In order to fully appreciate the sustainability of the SRP model, one needs to understand how and why the technical system implemented in Arizona responds to specific constraints at the intersection of ecology and economy (Mann, 1963; Kupel, 2003). As an initial step, it is important to examine the social framework dating back to the early twentieth century that generated such a utility (first for local agroindustry, then for urban growth) because it provides relevant and valuable insights into the role of various coalitions in the establishment of SRP in the region. Yet this historical perspective is not enough to capture the whole picture of this pioneering model and it must therefore be cross-linked to a sociological analysis of specific practices geared towards combining a large set of factors: standards of technical systems; contracting and pricing issues; and the environmental constraints for sustainable growth in a context where urban sprawl has been constantly increasing (Abbott, 1981, 1993).[2]

Box 2.1 The structure of the SRP

The SRP is not a usual business company – it refers to two separate entities:

> The Salt River Project Agricultural Improvement and Power District (the District) is an agricultural improvement district organized in 1937 under the laws of the State of Arizona.[3] It operates the Salt River Project (the Project), a federal reclamation project, under contracts with the Salt River Valley Water Users' Association (the Association), by which it has assumed the obligations and assets of the Association, including its obligations to the United States of America for the care, operation and maintenance of the Project.
>
> (SRP, 2014a: 6)

The SRP's Power District is a public utility and a political subdivision of Arizona whose board members are elected by landowners within its territory, owns and operates an electric system that generates, purchases and distributes energy, and provides electric service to residential, commercial, industrial, mining and agricultural users in the city of Phoenix and its neighbourhoods. The Association is a private water company that manages a system of dams and reservoirs; it is responsible for the construction, maintenance and operation of a supply system to deliver raw water for irrigation and municipal treatment purposes. The Association provides the water supply for an area within the major portions of the metropolitan area of Phoenix, its surrounding cities (including Glendale, Mesa, Tempe, Chandler and Scottsdale) and the Gila River Indian Community. It administers the water rights of SRP's 375-square-mile water service area and it operates and maintains the irrigation and drainage system. The members of the Association's Board of Governors are also elected by the landowners within the water service territory.

Operating revenue from retail electric and other electric was $2.6 billion in 2014, whereas the Association declared $14 million of benefits in 2014, for a covered population of 983,745 users (SRP, 2014a: 27). With a profit of $14.4 million per capita/year, it is no exaggeration to say that the water service is almost free, and that the costs of the service are covered by the sales of electricity. The interdependence of the Association and the Power District allows such a low-cost distribution, but the declared investments of electricity benefits into SRP activities are $62,184,000 ($6.32 per capita/year).

Water for a New America: the SRP model

The Reclamation Movement (1903–1937)

The first federal water policy was introduced in the USA with the Reclamation Act voted by Congress in 1902 (Pisani, 1982, 2002; Cortinas *et al.*, 2015). This legislation has been promoted and steered by agro-industrial enterprises of California, bankers of San Francisco and business lawyers well connected with politicians in Washington (Reisner, 1985; Gottlieb, 1988). Their objective was to address problems concerning the distribution of water in order to support economic growth. Originally, most of the western land was in the hands of the state, but some private lands were owned by railway companies, banks and other large landowners who sold it to farmers (Pisani, 1984). The fall in the price of farms however destroyed the confidence supporting the speculative market of agricultural products (Hofstadter, 1955: 57). As a consequence, land prices dropped in an unprecedented manner, people left the country for the city in droves, and many farmers were unable to repay their debts. The region's economic forces began to promote federal irrigation projects in order to support the local economy. They were also aware that another sector, namely mining, despite its profitability, could not ensure their continued expansion, and they viewed the arrival of new occupants as the most effective way of transforming the West into a new economic El Dorado (Hundley, 2001).

In 1897, a Californian water lawyer, George H. Maxwell, became the founder and leader of the National Irrigation Association, an institution that lobbied Congress on behalf of the advocates of federal irrigation projects (Pisani, 1994). He was supported by Francis G. Newlands, a landowner who became a congressman in 1893, and used his position to promote a federal irrigation policy. But beyond the economic goals, the content of the Reclamation Act was based on reformist ideals. Irrigation projects run by the Reclamation Service were linked by law to properties that were limited in size to 160 acres, in order to promote the development of small family farms. Furthermore, the law included measures designed to preclude speculation on land designated for irrigation projects (Hays, 1999). A water management system was also created by the users of these irrigation projects. The system was made up of districts responsible for forming associations to regulate irrigation within the geographical limits of water basins. There was a consensus about the need to develop a model of economic expansion capable of resolving the economic crisis that had ravaged the country in the late nineteenth century. During this initial period however, very few projects were completed. In 1920, only one major project was actually completed, namely the Arizona Salt River Project (Summit, 2013). But if the impact of the Reclamation Service through the 1920s was questioned, it had a powerful effect on the Phoenix area.

Indeed, the National Reclamation Act included a provision for landowners related to federal assistance for dams and irrigations works, which also applied to those living in the Salt River Valley around Phoenix. In Maricopa County, the construction of a dam and a reservoir on the Salt River, located 65 miles northeast of Phoenix, was promoted by investors of the city that were heard in Washington (Zarbin, 1986). To be qualified for this federal assistance, landowners, namely farmers and ranchers, were obliged to create a water users' association, which would have contractual liabilities to the government, such as buying up canals that were owned by the Arizona Water Company, a private company, settling the issue of water rights, and ensuring the complete reimbursement of building work to the government. The Salt River Project therefore paved the way for the numerous multi-purpose projects to come, which were not only designed to irrigate arid lands but also to electrify the country and thereby contribute to regional development. The construction of water reservoirs went hand in hand with the construction of power plants, which not only had the function of responding to the increasing demand for electricity from small municipalities around Phoenix, but also the financial function of allowing the federal government to be reimbursed for the cost of the dam through the sale of electricity to taxpayers.

The Salt River Valley Water Users' Association and Roosevelt Dam

Arizona Water Company, a New York-based private company, was the owner of the surrounding canals through which water was sold to farmers and ranchers. This company had a contractual relationship with the Pacific Gas and Electric Company, a California company, which resold electricity generated by the water from the dams to inhabitants of Phoenix. Farmers and ranchers had a strained relationship with the Arizona Water Company as they were dependent upon it for water to irrigate their land. By the early 1880s, the construction of a dam along with the creation of a reservoir was considered to be key to economic growth in the Salt River Valley. The problem was that the price of such a dam was very high, estimated at between $2 and $5 million; and the USA forbade the territory of Arizona, which was still not part of the Union, to incur such a debt (Fudala, 2001). Despite the advanced irrigation system enjoyed by the Valley, however, an assured water supply was still critically important for growing crops and raising livestock owing to the repeated droughts that landowners endured.

For over a decade, private companies and local communities tried unsuccessfully to finance a dam. A special event sped up the process for change. At the time, a national movement for local water recycling projects funded by the Federal Government was developing (Carpenter, 2001; Pisani, 2002). The lawyer George H. Maxwell became the leader of an alliance of economic interests and promoters of federal irrigation projects in the existing states of the time. In Phoenix, the businessman Benjamin A. Fowler quickly assumed leadership among local stakeholders. Beginning in 1901, Fowler went to Washington, DC to represent the interests of the Salt River Valley and to lobby Congress on behalf of Arizona. He worked along with Maxwell, the Republican Francis G. Newlands of Nevada and Senator Henry Hansbrough of North Dakota on a bill that would provide a way for local organizations to create a pool of water as well as to implement projects for supply services (Shermer, 2013). The bill won the support of Theodore Roosevelt, for whom the future of the West could not be separated from the management of its natural resources, including water. Roosevelt signed the National Reclamation Act on 17 June 1902.

The SRP did not exactly fit with the National Irrigation Act guidelines of the Reclamation Act. Indeed, in the Salt River Valley, all the lands potentially eligible for federal aid belonged to the private sector. The National Irrigation Act's purpose was to make public land available to settlers while facilitating agricultural work by irrigation systems. Envisioning this difficulty, two Phoenicians who were also members of the National Reclamation Association successfully advocated for an amendment to the Act, allowing private landholders to jointly collateralize their lands to guarantee repayment of construction costs (Needham, 2014: 37). Through the passing of this amendment, additional motivation arose for the creation of the SRP. The choice of the Salt River Project was mainly driven by several factors: the perfect location for the dam, the organizational capacities of landowners and the unfailing support of the National Reclamation Association, led by Maxwell. In the end, the success of the project would give an impetus to the national irrigation movement (Zarbin, 1986: 44).

The Association was incorporated in February 1903. In 1904, the Home Affairs Secretary lent his support to the contract with the Association for building the Roosevelt Dam. Before obtaining the needed federal loan, members of the Association had to use their land as collateral to prove to the Department of the Interior that the reservoir and dam construction costs would be repaid. The landowners of over 200,000 acres in the Salt River Valley formed the backbone of the Association where the share of holders varied according to the size of land owned. The Association took the form of an "unnamed company", which included landowners in the decision-making process of irrigation plans. The major representatives from the Association (the Board of Governors) were elected by an electorate of owners who had committed their land values into the irrigation project.[4] The businessman Benjamin A. Fowler, who had defended the project in Washington, became the first president of the Association. Shortly after, the dam on the Tonton Basin site became one of the first five projects to be authorized under the Reclamation Act.

In 1905, a major flood destroyed a dam that had been built by the Arizona Water Company. The management of this private company was hence increasingly criticized, leading to a new series of negotiations between water users and representatives of the federal government. This natural disaster accelerated the public acquisition process of a canal system which was initially owned by this private society. The repair of the dam was too expensive and water users wanted a solid cement construction. As chairman of the Association, Fowler left for Washington to negotiate the purchase of this operating canal system with the Department of the Interior. In 1905, the Reclamation Service decided that the ownership of the Arizona Water Company should be acquired before the dam construction began. The Reclamation Service officially took control of the property of the Arizona Water Company and of its canal network. The acquisition of the Arizona Water Company property gave the Federal Government a right of way on which to start work.

Nevertheless, for Washington, the implementation of the project was still a problem, as the prior water rights remained to be settled. Where there would be disputes over water rights, the government would incur no expense. According to Washington, a court had to identify what land would be irrigated to determine the amount of water available to be distributed to the land selected for refunding the cost of infrastructures. The problem was that, in the Salt River Valley, farmers were fighting among themselves and with the owners of canals including the Arizona Water Company to settle who had the right to water and the amount they could use. The attorney of the Association filed a lawsuit in the Arizona Superior Court in Maricopa County to adjudicate the rights of all people using or diverting the water flow of the Salt River, and to select complaints from water users asserting the oldest priority claims. In March 1910, the verdict of the trial led to the Kent Decree, which determined that almost 240,000 irrigable acres in the Salt River Valley had a right to water for agricultural purposes (Zarbin, 1986: 113–114). The Kent

Decree draws upon early principles of prior appropriation whereby the party who was first able to beneficially use the resource held the rights to it. Importantly, it allowed for the land to be seen has having value only when there was water associated with it. The right to the appropriation of water was based on beneficial use (prior appropriation doctrine), and the right of appropriation depended upon a supply of water (as a common good). The Kent Decree regulated the right to water for more than 200,000 acres in the Salt River Valley, which are still applied by SRP managers in the 2010s in the absence of the completion of the ongoing Gila River Adjudication.

A "circular" model: water and energy engines for economic growth

The Salt River Project Agricultural Improvement and Power District

During the building phase, the US Reclamation Service was responsible for the operation and maintenance of irrigation facilities. Landowners, members of the Association, were never satisfied with the operating procedures of the US Reclamation Service. A dispute over management issues broke out on 1 November 1917. Later, the US Reclamation Service handed the development, operation and maintenance of the irrigation project to the Association, which thus became the administrator as well as the private manager of the water system and electricity (Smith, 1972: 16). In 1917, the Association continued to operate the dam and the canal structures while the federal government retained ownership. The rights of the stored and provided water by the system belonged to the owners of the Association. By 1917, the Association had full responsibility for the Salt River Project. Water was supplied to 4,800 customers in the Valley, and the electricity from the Roosevelt dam was sold wholesale to 49 customers.

In the 1920s, the Association built a rural distribution grid and two hydroelectric dams to increase the production capacity and integrate more irrigated lands and thus persuade more members to join the project. The construction of power lines was ensured by a contract with the Central Arizona Light and Power Company (CALAPCO), a private company under contract with the City of Phoenix since the nineteenth century (Needham, 2014). In 1928, SRP and CALAPCO signed an operating agreement establishing the territory where each would provide electrical services and CALAPCO agreed to serve no more than 15 per cent of the customers on land owned by members of the Association (Needham, 2014: 39). Originally, the Salt River Project provided only the industrial sector (copper mines in the northeast of Phoenix) with water. Since 1928, the Association provided power to farms owned by the shareholders of the Association. This development was reflected in the electrification of rural areas nearly ten years before the national rural Electrification Act was passed as part of President Roosevelt's New Deal. Ninety per cent of the farms included in the Association had electrical services, representing, SRP bragged, "the largest block of electrified farmland in the world" (Shadegg, 1969: 14). By the early 1930s, in the context of the Great Depression, the Association found it hard to repay debts generated by the construction of the Salt River Project. Electricity sales dropped as copper mines closed. Farmers lost an important market for the sale of their products and were no longer able to pay the Association for water. The main sources of income of the Association were threatening to run out if new rules were not adopted.

From the 1930s under the Roosevelt administration, the Federal government's water policies entered a new phase in terms of objectives, resources and beneficiaries. Multi-purpose reclamation projects were launched in the country: they would provide hydroelectric power, deliver water and give flood protection to Western cities, such as Phoenix. This period was thus characterized by an increase in the number of water projects. From 1902 to 1933, the average annual budget for the Bureau of Reclamation's projects was $9 million and it completed 36 irrigation projects, but it raised $52 million and finished more than 200 technical systems (reservoirs, lakes, canals, dams) between 1933 and 1940 (Reisner, 1985: 165). In terms of water policy, the implementation of large hydroelectric plants that also controlled the flow of the great American rivers enabled several states to benefit from a single project and, more generally, the actions of the Federal government to overcome the economic and social consequences of the Depression brought more interdependence between states (Pisani, 2002: 252). Finally, this period was characterized by the rise of agribusiness and the cities, which were the leading beneficiaries of water projects, over and above the initial projects of the water policy designed to favour small farmers.

To help the Association get through the crisis, Arizona-elected officials allowed the establishment of agricultural districts in 1936. Their formation led to the creation of the second entity included under the umbrella of the SRP (the Salt River Project Agricultural Improvement and Power District). According to Arizona state law, the District was granted the rights, privileges, exemptions and immunities of a political subdivision of the state. The main reason for creating this new entity was to refinance debts incurred through the construction of facilities built during the 1920s with tax-exempt municipal bonds carrying a lower rate of interest repayment. Through these municipal bonds, the SRP was able to finance the construction of coal-fired generation plants to produce more electricity. Sixty years later, power operations would become the major business activity of the SRP, producing more than 90 per cent of its annual revenues (G. Gammage, interview, 2015).

The main change was that all the property included in the Association was transferred to the District. The administrative organization exploiting both water and electricity systems always structured the Association, but it was split into two legal entities: the Water Users' Association (private company) and the Power District (created by state law). As the District boundaries were adjusted to match the territory of the Association, the voting rights were given to owners in the District on the basis of one vote per acre. The theory, still true today, was that those with the most property would bear most of the responsibility for debt and risk, and should therefore have the greatest say in the governance of the District. The Association continued to manage water while the District assumed authority to sell electricity. Revenues from electricity allowed the Association to offer very low-cost water (C. Ester, interview, 2015). Indeed, the Association applied the principle of the Reclamation Act: "Power is the paying partner of water" (Smith 1972: 47). In fact the water and electricity activities of the Salt River Project were complementary in that they sought the same goal: reducing the operation costs of water supply.

Impacts of pro-growth coalitions upon the SRP

In the second half of the twentieth century, Phoenix and Tucson experienced urban expansion triggered by a quest for natural resources that were not available on site (Mohl, 1990; Rice and Bernard, 1984; Schipper, 2008). After the Second World War, the two biggest cities in the Sun Corridor experienced a high rate of expansion. Between 1945 and 1960, the population of Phoenix grew from 65,000 to 439,000, while Tucson's leapt from 38,000 to 212,000. Besides the local promotion of the defence industry and new technologies, growth focused on tourism and services, whose comparative advantages – a sunny climate and a cactus-filled desert – were part of the Western Ranching Lifestyle of the cities of the American West (Gober, 2006; Logan, 2006). Real estate became a prime component of the Sun Corridor's economy (Molotch and Logan, 1987). Land purchases (investment) and exploitation (construction) – farmland was frequently used for the purposes of urban development – generated a substantial economic windfall: "Fortunes in post-war Arizona were not made in gold or silver mining but in real estate. The state became a vast monopoly of new lands waiting to be developed" (Sheridan, 2012: 353).

Not everything went smoothly, however. A conflict between the City of Phoenix and the Salt River Project had broken out in the 1940s on the interpretation of water law. According to the city, the landowners who had converted their land from agricultural to residential, commercial and industrial use had abandoned their water. Their water rights were about to be subject to reappropriation. The city of Phoenix and other cities in the Salt River Valley had applied for water rights. The Salt River Project argued that water and land were tied together, independent of residential use. The answer to the question "Whose water is it?" was becoming crucial in that it would determine the new water allocation arising from increasing urbanization (Smith, 1972: 79). Up until then, the SRP had provided water and electricity to the rural parts of the Salt River Valley. The service area for irrigated water included 240,000 acres within a workable area for agriculture. Phoenix and other cities in the service area of the Salt River Project accounted for less than 10,000 acres and were being supplied with drinking water from their own wells. In Phoenix, many shareholders of the Association (including individuals and small farmers) also received irrigated water (Keane, 1991). From 1937 to 1941, 97 per cent of electricity was distributed to copper mines, other service providers, public authorities and used for groundwater pumping. The Association met the demands of users who consumed a lot of electricity rather than individual applications.

The Second World War transformed Central Arizona and the Phoenix area. The copper mines again prospered because the war effort required metal use. The SRP and the Phelps Dodge Mine Company cooperated to build the Horseshoe Dam on the Verde River to secure water for the exploitation of copper. After the war, professionals from aerial military operations who came to be trained in Arizona and had enjoyed the good weather in the area returned to take up a career and start a family. In 1947, the initiation of operations by Motorola in Phoenix contributed to growth in the region. In the 1960s and 1970s, Motorola became the largest private employer and the largest customer of SRP. With the introduction of air-conditioning in residential areas, Phoenix became livable throughout the year. This change also contributed to the increase in tourism, which had long been an important factor in the economic prosperity of Arizona. Increasingly, new residential developments and increased industrial and commercial activities replaced the farms (Ross, 2011). Such growth in the distribution territory of the SRP meant that the needs of the Salt River Valley for electricity and water were increasing, despite the decline of agriculture. The SRP had begun as an ancillary benefit of the dams and quickly, through power operations, became a major economic engine.

Entrepreneurs and heads of companies affiliated with the Chamber of Commerce and, supported by numerous local councils, also influenced urban and environmental policy by supporting tax rates and regulations intended to encourage major infrastructure projects. Business leaders knowingly used local government to encourage economic growth and to benefit from the financial manna provided by the residential and commercial construction industry (Shermer, 2013: 6). The increasing number of "shadow governments" – agricultural corporations, interest-based associations, committees, districts, homeowners and resident associations – in all sectors of social life (Garreau, 1991; Guttman and Willner, 1976), under cover of an alleged plurality of power centres, reinforced the pro-growth coalition's decision-making capacity and influenced city governments and the state legislature. This pro-growth coalition, which received little criticism, promoted the urban expansion of Phoenix and Tucson through a proactive policy designed to encourage economic development. Within this context, SRP has successfully cultivated a public image as a beneficial "shadow government" (Gammage, 1999: 25), delivering public resources of water and power and managing a vast watershed for the benefit of the Phoenix metropolitan area.

Towards promoting a conservationist approach to water and energy

Legal and technical responses to urbanization

In 1952, the Association signed the Domestic Water Contract with the city of Phoenix. Initially, the Association supplied the "Farm water user" which designated a variety of agricultural users specializing in cotton, grain and horticulture. Irrigated land was used as pasture for dairy cattle and fodder crops were used for cattle feed. Just before the depression of the 1930s, water was distributed in residential subdivisions and municipalities, for irrigation purposes, as well as for farms. The Domestic Water Contract gave birth to a new category of user: the domestic water user. This user was dependent on the ability of the Association to provide water, but had no direct contact with the Association as the city was providing the user's water (Smith, 1972). After the Second World War, electricity consumption was not only covered by agricultural activities, but also by industrial, commercial and residential uses (these activities increased from 3 per cent of total consumption for the period 1937 to 1941 to over 60 per cent for 1967).

To adapt to this change, the electricity distribution from the Salt River Project had to take into account the growing needs of individual consumers. The Salt River Project was not the only organization to provide electricity. The major competitor of the District was the Arizona Public Service Company, which, despite its name, is a private company operating within the same territory served by the SRP. The territorial agreement from 1955 aimed to delimitate electrical territories for the District and Arizona Public Service Company, moving competition between the two organizations to new high technologies and residential developments in their distribution areas. The rate structure, which was greatly advantageous for customers, was one of many factors that could generate more economic growth. The two competitors shared the same goal, namely supplying electricity at low cost to the Valley. But the rapid rate of urban growth led to ever more new electricity customers and thus more new consumption habits (Scott and Pasqualetti, 2010). Given this context, competition increased, affecting the relationship between the two electricity suppliers.

Faced with urbanization and increased competition, the SRP undertook to rehabilitate water supply systems and electrical transmission dating back 50 years before in order to meet the new growing needs of urbanization. The SRP built three steam-generating plants in the Phoenix area to serve its growing territory. For this reason, among others, the SRP supported the implementation of the Central Arizona Project (August 1999), a 360-mile-long canal supplying Colorado River water to the southern areas of the state, which would provide another important source of surface water to Arizona while thereby relieving water stress to Phoenix (Johnson, 1977; Coeurdray *et al.*, 2015). The determination of the area, which could be irrigated, was still problematic in 1967. The urbanization of agricultural land had stimulated requests for urban water. The cities within the irrigation project had the responsibility of meeting the new urban demand (i.e. urban domestic water demand). They therefore turned to the Association for additional water supplies. The rule of the Association stated that cities could only use water for irrigation on urbanized land for which appropriations for water rights had been executed. But cities defended the fact that they had to have these lands for water use in any part of the municipal water system, whether within or outside the borders of the Salt River Project. The Association had been created to ensure that landowners, who had put up their land as collateral for the Salt River Project, would receive an adequate supply of water for irrigation of this land. The water supply is tied to irrigated land. The territory for which the Association was responsible included the lands of members of the Association within the Salt River Reservoir District.

With urbanization and population growth, the water customer structure of the Association would change, by becoming more heterogeneous. It was the same with electricity. Initially, customers were mining and other infrastructures. Later, the power lines were extended to include groundwater pumping stations and farm families. In 1949, 18,513 electricity customers were called "rural domestic customers". In 1967, the number of residential customers had increased to 128,966. From the 1970s onward, the SRP was a leading partner in several regional projects of power plants fuelled by coal. The SRP built and exploited the Navajo Generating Station in Page and built another coal power plant, the Coronado Generating Station. The SRP also joined the Nuclear Arizona Resource Study Group, which would assess the costs, resource needs, the potential risk, the size of facility and siting issues for what would become the Palo Verde Nuclear Generation Station. Finally, the SRP secured a 17 per cent share of the plant that is operated by the Arizona Public Service, which also owns a share of it. Since 1980, the SRP has provided almost 13 trillion kilowatt hours to 357,000 customers.

Competition for a scarce resource: expanding suburban cities and conservation policy

Suburban expansion in the Sun Corridor based on "unfettered growth" has gradually been called into question due to the emergence of new environmental constraints imposed by the Federal Government, especially upon the protection of Arizona groundwater (Avery *et al.*, 2007; Vincent, 2006). The continued expansion of Phoenix and Tucson caused greater competition for water also desired by farmers, mining companies and municipalities that did not operate in the same hydrological regime. The ideology of growth, on behalf of which leaders in Arizona (senators, representatives of farming and urban interests) had been lobbying Washington for over 30 years to guarantee the economic development of the region, was adjusted to a new environmental situation. A new coalition largely made up of politicians then began to encourage a change in policy with a view to encompassing the protection of aquifers, a *sine qua non* for the Central Arizona Project (CAP) to provide water to southern Arizona (Coeurdray *et al.*, 2015). More generally, the development of water infrastructure has been impacting the economic development of Arizona throughout the twentieth century up until this day, as shown in the example of the City of Scottsdale (see Box 2.2).

Box 2.2 Water infrastructure and economic development

Located east of Phoenix, the suburban area of Scottsdale has been allowed, beginning in the early 1900s, through the Salt River Valley Water Users' Association, to implement irrigation and recycling water projects, with the building of two dams in order to guarantee a water supply all year round to support the agricultural economy of the Valley and Scottsdale. The dams were an important improvement in the management of the Salt River in that farmers were no longer dependent on temporary dams, which were often destroyed during floods caused by heavy winter and summer rains. With these new resources and irrigation, the culture of cotton took root in the Valley. The farmers devoted more and more large tracts of land to grow this crop. During the First World War, when the cotton supply from other countries was embargoed, demand for long staple cotton increased (Schulman, 1991). With the increase in demand for cotton, farmers needed manual labour. The least restrictive immigration policies with Mexico allowed farm owners to bring in Mexican families to work in their fields (Fudala, 2001). Scottsdale's booming economy along with more and more farms, ranches and businesses required a growing supply of water. In addition, the success of pioneer families between the 1900s and the 1950s drew newcomers to cultivate productive land in northern Scottsdale and raise livestock. In this context, the Association continued to expand, by building more dams and reservoirs. By managing the Arizona Canal which runs through Scottsdale, the Association contributed to the growth of this city of pioneers and, at the same time, the city of Scottsdale contributed to the promotion of the Association. Apart from water, the Association supported the boom of Scottsdale through cultural activities and by participating in events and charitable organizations. Through Scottsdale, the SRP shows the twin aspects of its role: maintaining the pioneering spirit, with a sense of innovation, partnership and community, while practising performance art with its leading position in the water and energy industries.

Scottsdale is a good example of how a water source that is viewed as reliable and sustainable can impact the growth and prosperity of a desert community. The collaboration of Scottsdale and Phoenix to provide a safe and reliable water supply was on full display in October 2015 at an SRP-sponsored event in Scottsdale entitled "Vibrant Forests, Vibrant Economy", with the subtitle "Finding Solutions". At this conference the Mayor of Phoenix, Greg Stanton, spoke eloquently on behalf of his city by citing the importance of managing a diverse water supply portfolio. He was also unabashed about speaking about the significance of recognizing climate change, which allowed him to link the importance of forest health, the SRP, his city and the cities around Phoenix under a unified call to action to promote forest and water supply health. The Mayor of Scottsdale, Jim Lane, had previously opened the conference with a talk entitled "Why Forests Matter to Desert Cities", emphasizing the importance of watershed health to the vitality of the metropolitan area. This concern to preserve watershed by forest conservation is not new (Hayes, 1999), but in the context of drought, it acquires a new importance (C. Ester, interview, 2015).

Both the Association and the City of Scottsdale henceforth assume a significant role in championing the conservationist attitude towards water and energy management, by promoting professional practices that combine more visibly ecological concerns and economical orientations. The SRP's communication strategy is to promote actions taken since the beginning of the 2000s towards forest protection against ravaging fires that destroyed watersheds and biomass resources, the aim of which is to secure sustainable water and energy systems. Far from being an epiphenomenon or a single marketing campaign for new customers, this high-profile strategy expresses the awareness of shifting challenges faced by the SRP, such as urban sprawl and climate change. While, in the 1970s, the Federal Government had cut funding for major water systems and imposed a set of environmental standards upon groundwater in the region, the SRP did not seem overly concerned about the pressure. Only since the 1990s did SRP formally take note of the new environmental constraints, at times when the construction of the CAP facilitated the economic development of Arizona while enabling it to think about water management in a more global way.

In the 1990s, shortly after the arrival of CAP water in the region, the SRP created GRUSP (Granite Reef Underground Storage) which is regarded by the SRP as the first major recharge facility in the state and one of the largest in the nation (SRP, 2014c). Water stored at GRUSP primarily comes from the Central Arizona Project aqueduct, which is connected to the SRP's canal system. Since 1994, the Central Arizona Water Conservation District, which operates CAP, has been working with the SRP to coordinate water deliveries to GRUSP. From 2000 onward, the SRP's strategy has been increasingly geared towards establishing partnerships with other states who share similar concerns over water management in the region. Given the steadily decreasing level of the Colorado River (Mount *et al.*, 2015), Arizona and Mexico seek to anticipate the potential risks of water shortages, knowing that California is not inclined to renegotiate its share of the Colorado River, which was laid down by the Colorado River Compact in 1922. In May 2014, as part of the Arizona–Mexico commission, representatives from Arizona (the Arizona Department of Water Resources, SRP and CAP) and from Sonora (Comisión Estatal del Agua and Comisión de Cooperación Ecológica Fronteriza) shared experiences about resolving crucial issues with regard to local and regional water supplies, and discussed water rights and economic development. According to Marc Campbell, SRP Senior Water Rights and Contract Analyst, this meeting was necessary because of the serious challenges experienced by both states: "The development of additional water resources, infrastructure and amenities is critical to the future of Arizona and Sonora" and "both states have a growing need for more supplies to either improve existing services or to keep pace with increasing populations and anticipated economic development" (SRP, 2014c). Desalination for an additional water source to increase water supplies was among the debated topics. The Mexican party showed its interest in building a desalination facility site on the Gulf of California (on the Sea of Cortez). Since then, the SRP has been considering being a financial partner on a pilot project that may benefit both states for this type of project. Indeed, for the SRP, which takes into account the repeated droughts that have been impacting water supplies in the region, getting involved in such a project would also be a means of negotiating with Mexico over its own Colorado River share in case of water shortages in Arizona (C. Ester, interview, 2015).

In the broader context of climate change, the Environmental Protection Agency now requires the implementation of new energy standards for states, including Arizona. This contrasts with the practices in the past. According to the Reclamation Movement in the early twentieth century and the New Deal policies, huge dams would be used to supply agriculture and electrify the country. In the 1970s, under the pressure of environmental movements, these mega-hydro-dams were deeply criticized because of their impacts upon ecosystems. Political battles focused on the preservation of the Grand Canyon, regarded as the environmental jewel in the West and not as a potential economic asset. Instead of hydro-dams, coal-fired power plants as new sources of energy were promoted for a less damaging use of natural resources. Since the 2000s, with the debate over reducing greenhouse emissions, the use of coal as a source of electrical energy has become increasingly challenged. Coal is now seen as a dirty energy. In Arizona, this change in perception jeopardizes the way in which energy and water policies have been implemented since the arrival of CAP. Operated by the SRP, Navajo Generation Station (NGS) is a 2,250-megawatt coal-fired power plant located just outside of Page on the Navajo Reservation. NGS is critical to the Central Arizona Project, since more than 90 per cent of the energy used to power CAP comes from the generating station. Through pumping plants driven by coal-fired turbines, CAP is able to bring about 1.5 million acre-feet of Colorado River water per year to Pima, Pinal and Maricopa counties. Faced with threats of closure of NGS by the Environmental Protection Agency (EPA), a Technical Work Group (TWG) was formed to find a way of identifying alternatives to reduce emissions. This group includes representatives from the Central Arizona Water Conservation District, which operates CAP, the Environment Defence Fund, the Gila River Indian Community, the Navajo Nation, SRP, the US

Department of the Interior and Western Resource Advocates. An agreement was reached in October 2013. The TWG aims to achieve greater emission reductions than the earlier proposal from the US EPA, which, had it been implemented, would have required such costly controls for emissions at NGS that CAP water would have become unaffordable for its customers and furthermore compromised an additional source of water for the SRP.

Conclusion

This new situation highlights changes occurring in large technical systems in charge of water and energy management in the West. We see here how both water stress in a regional area and global environmental pressures can have an effect on local provider practices, just as these very practices can contribute to redefining national and international orientations. This new situation sheds light on the position of the SRP. More frequently presented as a pioneering and unusual model, the SRP has increasingly become part of a wider and more complex system of natural resources management in which various players interact with each other. In order to avoid jeopardizing their own practices, they are forced to adapt to a new context by seeking innovative solutions, which combine legal rules, public interest, economic performance and increasing ecological concerns. This attitude oriented towards water conservation would explain the reason why the SRP can be both a protector of floods and forest, listen to new controversial technologies such as desalination and inclined to be a clean energy provider while promoting local and regional economic growth.

The sociohistory of the SRP demonstrates the extent to which the legal categories of an organization (public or private entity) that are commonly used to define a technical system prove to be irrelevant in order to grasp the water infrastructures development as a whole in the southwestern USA. By focusing on the ways in which water management models have continually been resilient to exterior factors, such as growing urbanization, while providing innovative ways to adapt to them, this chapter has revealed how the social forces at work in water policy are embodied by institutions capable both of reconciling the various interests involved and generating a multifaceted spectrum of professional practices. By depicting the institutional aspects of the SRP's history, it became possible to show how public and private interests overlap and highlight the multilayered decision-making process (with respect to specific locations ranging from land plot and water basin up to metropolitan areas) on various action plans (such as concluding contracts, establishing innovative financial schemes, agreement of sharing land for calculating the appropriation of water and so on). It also enabled us to integrate these elements into a sociological analysis framework in which they play out as explanatory factors of professional and political approaches towards dealing with water issues. Bringing back the history into such a sociological perspective showed that water projects cannot only be considered as a matter of technical ability or the product of rational decision-making but should be regarded as deeply embedded in formal institutions and long-standing efforts to adapt practices to changing environmental conditions. Thus one can better appreciate what the SRP's motto ("Delivering more than water") finally means.

Notes

1 See Part II of this volume.
2 This research is part of a broader sociological investigation into the field of water policy in the western USA. It is based on interviews with water managers from Arizona and on an analysis of documents and data related to the implementation of water services from the 1890s to the 2010s.
3 For a history of the Salt River Project see: www.srpnet.com/about/history/water.aspx. For a presentation of the Salt River Project's water rights see East Valley Water Forum: http://evwf.org/wp-content/uploads/2015/01/AppendixC-WaterSuppliesAvailable.pdf.
4 The legal entity of the Association was initially designed to be both a company (corporation) and a cooperative (Smith, 1972: 15). Then the Salt River Valley Water Users' Association was designated to represent owners to the government.

Part II

Questions of legitimacy

Private interventions in a public service

3 Legitimacy in the test of time

Operating water in Buenos Aires

Bernard de Gouvello

The concession for the management of the water and wastewater services for the city of Buenos Aires was contracted out in April 1993, and officially cancelled by the Argentinian state in March 2006.[1] This emblematic case has been the subject of extensive research and a plethora of studies by international bodies (Gaggero *et al.*, 1992; Faudry, 1997; Alcazar *et al.*, 2000; de Gouvello, 2001; Urcelay, 2007; Azpiazu, 2010). The explanation generally given for this reversal is the major economic crisis of late 2001, which ended the parity between the peso and the US dollar and disrupted the financial balance of the contract. The approach taken here is to show that the failure of the contract reflects as well a more profound and gradual process of loss of legitimacy on the part of the private operator, which deepened steadily throughout the operation, with several contributory factors: the expectations of the population, the behaviour of institutions, changes in the economic and political context, and finally certain positions adopted by the multinational company. This approach prompts a different reading of the history of the Buenos Aires water concession and demonstrates that the final result of contracting out is not only determined by the quality of the legal framework but also by the relation between the operator, the elected officials and "civil society".

Background and signature of the contract

In the early 1990s, the Buenos Aires water and wastewater services were managed by the public utility Obras Sanitarias de la Nación (OSN). Created by the Argentinian state in the early twentieth century with public health objectives, financed by the country's agricultural revenues, the company developed rapidly up until the 1940s, taking charge of the water and wastewater systems for the country's larger cities. From the 1950s onward, OSN underwent a crisis because of a drop in public funding and charges that were no longer sufficient to cover day-to-day operating costs. The crisis deepened in the 1970s as the services ceased to be able to keep pace with demographic growth. Formerly a flagship company, OSN became a burden.

In 1980, the Federal government transferred the services to the provinces: OSN's coverage was reduced to the federal capital and to the 13 surrounding municipalities (*partidos*). Nonetheless, services continued to deteriorate. Elected as president in 1989 against a background of hyperinflation, Carlos Menem immediately initiated state reform through Emergency Act No. 23.696/89.[2] A massive privatization programme for industrial activities and public services was introduced. However, the "privatization" of OSN did not take the form of a sale of the company's assets, but instead a concession system intended to remedy the service failures (supply outages and problems of water quality),[3] to extend services to the city outskirts as yet without water supply and to improve the company's productivity, while maintaining low tariffs.

In 1991, the government developed a strategy to attract the most famous water companies. Up until April 1993, prices were readjusted several times, which led to a cumulative increase that was significantly higher than the retail price index (Azpiazu, 2010). A technical committee was set up to analyse the techno-economic feasibility of the concession, to create a regulatory framework, to establish the minimum investment needed, to prepare the tender specifications and to monitor the process. The tendering process began in December 1991. A shortlist was established, following prequalification based on preset criteria (size, activity and financial soundness). Five large international companies were shortlisted: Canal Isabel II, Compagnie Générale des Eaux, Lyonnaise des Eaux, North West Water and Thames Water. The bids consisted of a technical component, which was assessed first, then an economic-financial proposal. Three bids were selected on technical criteria; the choice was then made on the basis of a coefficient of reduction (K) below the existing tariff. With a K of 0.731, the Aguas Argentinas SA (AASA) consortium, headed by Lyonnaise,[4] won the contract, just ahead of Thames Water (0.739). The concession contract, signed on 28 April 1993, came into force on 1 May. The AASA consortium consisted of several shareholders: Lyonnaise (operator, 25.4 per cent of the capital); two international firms specializing in water (AGBAR and Générale des Eaux); two national groups (Sociedad Comercial del Plata (SCP) and Grupo Meller (GM)) with no background in the sector, and one leading bank of the country; the employees held 10 per cent of the capital, the legally required percentage. This selection procedure demanded a great deal of effort by the participants. For example, Lyonnaise sent a full-time team of between 30 and 50 people to the country for six months.

The Buenos Aires concession was a hybrid of the French model of contracting out (with a long-term contract, in this case 30 years), and the English model of regulation (by a specific and independent body). It involved simultaneously: the Federal government as the authority which applied the contract and owned the infrastructure (the operating assets would be returned to it at the end of the concession); AASA, the concession holder responsible for operating and maintaining the infrastructures as well as extending them within the concession zone; and ETOSS (*Ente Tripartito de Obras y Servicios Sanitarios*), the regulatory body, made up of 80 civil servants from different institutions – the Federal State, Buenos Aires Province, the Federal Capital and executives from the former Obras Sanitarias de la Nación.[5]

The management of the concession was covered by a framework of highly detailed documents divided into five-year periods: concession contract, regulatory framework, tender specifications and the bid submitted by the consortium. Targets were set for quality standards and infrastructure development in order to improve the service. At the end of each five-year period, there would be a renegotiation to readjust the targets for the following period. The investment plan submitted by the operator (see Table 3.1) revealed a significant financial effort at the start of the concession, with more than two-thirds of the investment concentrated in the first half.

The contract was based on one fundamental principle: prices would be set to cover the cost of operating and of extending services, and to maintain a certain rate of return. Certain clauses covered the reasons for any price increases: significant changes in cost indices; change in the parity between the Argentine peso and the US dollar;[6] and changes in the taxes imposed on the concession holder. Finally, the contract included a requirement for a capital contribution of US$120 million, as well as an unconditional commitment of US$200 million from banking institutions.

The assessment of the deterioration in public services in general and of the situation of water in particular was widely shared at the time by the media and by economic analysis (Natale, 1993). The Argentine press at the time was full of articles describing the situation of the public services (electricity, telephone, post, water and sewage) as catastrophic: the expression "al borde del colapso" (on the edge of collapse) was commonplace. The solution of bringing in an enterprise with reference points in the water sector to recover the city's situation had been carefully prepared by the government, which also drew upon the experience acquired in the first set of privatizations – union opposition to the privatization of the telephone system had led to army

Table 3.1 AASA investment plan

	Five-year period (in US$ millions)						*Total*
	1st	*2nd*	*3rd*	*4th*	*5th*	*6th*	
Drinking water plants	37	29	20	6	10	0	102
Lifting stations and underground "rios"	109	48	58	7	9	0	231
Water mains	327	246	172	193	136	127	1,201
Sewage networks	206	258	315	239	133	129	1,280
Main collectors and pumping stations	100	16	5	24	39	1	185
Water treatment plants	244	101	369	86	130	6	936
TOTAL	1,023	698	939	555	457	263	3,935

intervention (Pierbattisti, 2006). In the case of water, the assets were not privatized but contracted out, the unions ultimately supporting the procedure. The award of the contract went through without significant opposition. The operation received the imprimatur of the international institutions, which were keen to promote private sector involvement in the operation of water and sanitation services, and saw the Buenos Aires concession as a desirable model. When the AASA consortium took charge of the services, its legitimacy seemed unquestioned.

Start-up and adjustments (May 1993 to February 1997)

During the first year, the company restructured the inherited workforce and implemented measures to improve the service with the existing infrastructure. Some 50 expatriate staff members from Lyonnaise des Eaux were introduced into the operator in order to spread a new corporate culture. With union agreement and state support, a "voluntary redundancy" plan for OSN personnel reduced the number of transfers to the new company from 7,635 to 3,628 in January 1994. A new collective employment agreement was signed with the trade unions.

On a technical side, priority was placed on the reorganization of the drinking water plants and on increasing their production capacity. Operations were carried out on the network to reduce the high leak rates (47%). AASA signed a framework agreement for the management of its investment programme with a subsidiary of Lyonnaise and a British engineering company. An initial analysis of the existing infrastructure in 1993, based on 2000 measurement points, was used to optimize the investment plan. Results followed. Drinking water production capacity increased by 20 per cent, leak repair times were reduced from 180 to 72 hours and waiting time for users in cases of blocked drains from 20 to six days. Above all, the company achieved an important symbolic goal: 1993 saw the first summer for 15 years without significant water outages.

The year 1994 saw the start of substantial efforts concerning a better knowledge of the terrain, engineering works, internal reorganization and revenue growth. A database developed from 24,000 samples taken over an area of 8,000 km^2 produced results. The drinking water network proved to be more damaged than expected; the assumption of a per capita consumption rate of 200 l/d/h[7] adopted in the term of references and then in the bids was wrong, the real per capita consumption rate being closer to 500 l/d/h (Faudry, 1997: 119, 122). The lack of meters, the existence of numerous private swimming pools on the outskirts of Buenos Aires, the age of sanitary appliances (causing high consumption levels) and the population's lack of awareness of rational use of the resource may explain this gap. It therefore proved necessary to develop production rather than prioritizing wastewater management. The focus of investment went primarily to reinforcing the drinking water production plants and reinstating existing equipment. Network extension measures began in 1994 with more than 60 projects

(1,500 km installed in December 1995, above and beyond the contractual targets; incorporation of the 530,000 inhabitants of Quilmes into the concession, etc.). AASA also conducted an internal reorganization: the creation of a customer relations department, dividing the concession into four zones, each with its own operational directorate, a new billing management and collection system, staff training programme, and the establishment of an in-house management school, followed by a technical training centre. The operator also took measures to increase its revenues, converting 40,000 illegal users into regular customers in 1994. Then, with the help of aerial photographs taken across the concession area, the land registry data used for price setting were updated (see Appendix 3.1). The updated tariff was applied retrospectively from the start of the concession. Finalized in the course of 1996, this update led to a 30 per cent increase in the amounts billed.

In June 1994, following several new environmental requirements introduced to protect watercourses (*Riachuelo* and *Reconquista*) and the incorporation of new government projects (supplying the established population of the Buenos Aires slums; the plan to replace well water in Buenos Aires in order to limit nitrate pollution), the Economics Ministry was the instigator of a first amendment to the concession contract. In return, the company was allowed to increase its tariffs (ETOSS Resolution No. 81/94). Relative to the original terms of the contract, the K coefficient was increased by 13.5 per cent, the minimum connection charges were respectively increased by 83.7 per cent (drinking water) and 42 per cent (wastewater), and the associated infrastructure charges were increased by 38.5 per cent and 45.7 per cent, respectively.

During the course of this period, the company encountered its first difficulty with the population. In 1995, the People's Defender,[8] reacting to demands from users and NGOs, took legal action against a programme to install a single meter in collective buildings and condominiums on grounds of equity: under this view, the choice for metering failed to take account of each apartment's real consumption. The judge's ruling signalled the end of the programme and – more importantly – it instituted a trend towards the "judicialization" of the contract. But the major problem was the extension of the services under another main programme. The operator that had expanded the network began billing new customers for "infrastructure charges" during the second half of 1995. In 1996, these charges gave rise to a number of individual complaints and collective reaction, which prompted the national government to ask the company to suspend the programme. This point then came up on the agenda for renegotiation of the contract launched in February 1997 by the government (Decree 149/97).

Renegotiation of the contract (February 1997 to January 2001)

Only the two parties' signatories of the contract – the Federal state and AASA – took part in the renegotiation; the other stakeholders were not involved in the process. The state was represented by the Secretaría de Obras Públicas y Transporte (SOPyT) and the Secretaría de Recursos Naturales y Desarrollo Sustentable (SRNyDS). The objectives were multiple: to resolve the problem of the infrastructure charges; to make the Water and Sanitation Master Plans compatible with environmental management of the rivers, and to contribute "to better fulfilment of the objectives of the regulatory framework" (Decree 149/97, Article 3). Several solutions were to be explored: a change to the term of the concession; postponement of investment and rescheduling of extension works; introduction of new investment; modification of the economic and financial parameters of the concession, and access to the networks by new zones. Running over more than two and a half years, the renegotiation consisted of two phases: the question of the infrastructure charge, and wider reform of the contract.

In November 1997, a major modification was introduced in the financing of the extensions (Decree 1.167/97). The "infrastructure charge", previously applicable only to new connections, was replaced by a charge applicable to all subscribers. This two-monthly amount of Ar$3.99 was dedicated to fund the cost of the network and of new connections for the drinking water and wastewater in the concession's extension zones, and to cover financing requirements for additional investments introduced into the wastewater plan. New subscribers continued to be charged a bimonthly "connection fee" of Ar$4 for each service, payable for five years. The end of the first five-year period was shifted to 31 December 1998. A consultative committee was established; coordinated by the Natural Resources Secretariat, it was responsible for specifying the provisions of the renegotiation; its members consisted of representatives of the three relevant jurisdictions (Nation, City of Buenos Aires and Province of Buenos Aires).

The redefinition of the concession's economic regulation constituted the second phase of the renegotiation. The parties confirmed the concept of equilibrium: the concession would be in financial equilibrium if the tariff allowed the operator to generate sufficient revenue to cover the planned investments, the costs of operating the services and a fair return. The operator and the Natural Resources Secretariat established a mathematical model to identify the concession's equilibrium point at a given moment. The tariff review system was also revised: subsequently, the parties would have to agree on future investment obligations, whereas an automatic annual adjustment mechanism was established. The reasons for an extraordinary review were also specified: in particular, any change in the peso–dollar parity would be carried over to the tariff structure (Resolution 602/99). The principle used for the calculation and revision of the concession holder's rate of return was set by a special index.[9] This calculation was to be carried out at each five-year review and was to reflect the real cost of the capital invested over the period. Finally, in August 1999, the Exposición Financiera Neta Quinquenal (EFNQ)[10] was approved, along with its application to the first five years of the concession (Resolution 602/99). On 24 November 1999, Decree

1369/99 completed the renegotiation of the tariff aspects, taking into account the amount of the EFNQ (a deficit of Ar$823 million, corresponding to the difference between investment and revenues). The questions relating to the period 1993 to 1998 were incorporated into the following five-year period (see Table 3.2).

Hardly had the contract renegotiation finished when discussions on the second five-year plan began for 1999 to 2001. In the meantime, the government had changed; signalling the end of the Carlos Menem era, Fernando de la Rua was elected as president in December 1999. In March 2000, AASA developed a draft plan for improvements and extensions of the services. In June, ETOSS held a public hearing so that the operator could set out its plan to civil society organizations, and the national, provincial and municipal authorities. In August, the government of Buenos Aires Province, the Ministry of Public Works and Services, the municipalities concerned and the regulator signed an agreement stipulating the priority to extend the networks and the water services to 1.5 million people on the outskirts of Greater Buenos Aires, most of them poor, as well as to the construction of several major facilities. On 9 January 2001, the second five-year plan was approved and the action plan confirmed: forecasts for extensions to the system were now defined in terms of the surface area supplied (130 km^2 for sewers, 78 km^2 for drinking water). With regard to the tariffs, it was decided that the coefficient K should be raised by 3.9 per cent for each of the years 2001 to 2003, in order to build up a specific fund (*fideicommis*) intended to finance part of the investment planned for 2001 to 2003. Finally, two social programmes were included (one for social tariffs, the other aimed at deprived areas), involving the state, the regulator, the company and also the municipalities.

Almost as soon as it was in place, this second five-year plan was undermined by the deterioration in the country's economic situation. At the beginning of 2001, Argentina was facing growing difficulties in paying its external debt. The economic and financial parameters on which this second plan was based changed, threatening its feasibility. On the one hand, the company was faced with increases in duties and taxes levied by central government, the

Table 3.2 Results of the first five-year period

	Start of the concession (1 May 1993)	*End of the first five-year period (Set at 31 December 1998)*
Drinking water supply	65%	79%
Sanitation supply	54%	61%
Customers	2,036,000	2,576,000
Extension to the drinking water system		+1,083 km
Extension to the sanitation system		+298 km
Total investment		US$1.170 billion

province of Buenos Aires and the municipalities. On the other hand, there was a significant rise in the number of late-paying customers. Finally, the deterioration in economic conditions prompted the rating agency Standard & Poor's to downgrade AASA's credit rating, with the result that the financial conditions of the company's existing loans became tougher. Between March and October 2001, AASA sent six letters to the application authority and the regulatory body (ETOSS) requesting an extraordinary review of prices, as provided for in the contract. On 19 October 2001, ETOSS replied, demanding additional information to assess the economic impact on the concession.

In parallel with the process of contract renegotiation and overhaul, AASA underwent a significant change in its shareholder structure. Between 1998 and 2000, the two Argentinian partners, Grupo Meller and SCP, had withdrawn from the consortium,[11] their shares being acquired by Suez Environnement (which succeeded Lyonnaise des Eaux) and AGBAR (40 per cent and 25 per cent of the capital, respectively). At the time, the concession was seen as a good deal, and the reinforced holdings made it easier to consolidate results in the respective groups. During this period, the socio-economic characteristics of the supply zones were taken into account, with the involvement of the municipalities. The system of regulation by contract, in its renegotiated form, seemed sustainable: knowledge of the territorial context and infrastructures, control of the production factors (operation and investment), and forecasts (for revenues, costs and investments) made on the basis of reliable data. However, conditions were changing: the party, led by Menem, a keen supporter of private companies, lost the 1999 presidential election, and the economic crisis began to deepen.

Crisis, negotiations and breakdown (January 2002 to March 2006)

At the end of 2001, as the economic situation was deeply deteriorating, in order to prevent rising capital outflows, the government froze bank accounts and imposed a weekly cap on cash withdrawals. Increasingly massive demonstrations prompted President La Rúa to resign on 20 December 2001. A week later, on 6 January 2002, the interim president Eduardo Duhalde put through a public emergency act which abolished the convertibility system (the parity between the peso and the US dollar): the Argentinian currency then floated to a rate of Ar$4 for US$1, stabilizing after a few weeks at a rate of around 3.4 for 1. The emergency act also decreed that for all public law contracts signed by the administration, the price and tariff indexation clauses would be established in Ar$ on the basis of an exchange rate of Ar$1 for US$1, that the state was authorized to renegotiate the contracts concerned and that, despite the freeze in tariffs, the companies should continue to fulfil all their undertakings. The economic balance of the contract was broken: with bills being issued in local currency, revenues expressed in US dollars were worth one-third of their original value, while the debt remained the same. Without an increase in prices the company could no longer meet its contractual obligations, or even repay its debts. AASA found itself obliged to start negotiations with both its lenders and the Argentinian state. It therefore refocused on continuity of service and suspended all debt repayments with the exception of a fraction of the interest payments: priority in the use of revenues was given to covering operating and maintenance costs. Inflation led to significant rises in production costs, while revenues did not increase because of the tariff freeze:[12] from Ar$/US$290 million in 2001, they rose to 347 million in 2002 and to more than 460 million in 2005.

From 2002, investment by the operator was limited to completing works begun in 2001 and to keeping services operating (emergencies on the secondary systems and preventive maintenance on the infrastructures). The social tariff programme provided for in the second five-year plan was nevertheless introduced, with Ar$4 million per year in the form of coupons being distributed to the municipalities, which allocated them to users on the basis of social criteria, a process monitored by the consumer associations and the regulator. During the period 2002 to 2005, more than 103,000 people consequently saw their bills fall by around 50 per cent of the minimum tariff.

Service extensions were confined to the introduction of water supply into deprived districts through two programmes. Devised by the operator, the *Modelo Participativo de Gestión* (MPG) began as a pilot in 2001, involving the operator, the municipalities and the populations concerned.[13] At the end of 2005, more than 100,000 inhabitants of Greater Buenos Aires had access to the drinking water supply through the MPG. However, Minister de Vido wanted a more massive programme of supply to deprived areas: he launched the *Agua + Trabajo* programme, based on a partnership between *Ente Nacional de Obras Hídricas y Saneamiento*,[14] AASA and the municipality of La Matanza, the inhabitants being represented by cooperatives.[15] By March 2006, around 350,000 people would have benefited from this plan.

By 31 December 2001, AASA's total debt stood at US$706.1 million (capital and interest). This debt consisted mainly of long-term loans, mostly from multilateral lenders: Inter-American Development Bank, International Finance Corporation, European Investment Bank (US$500 million) and ING Barings Bank (US$110 million). AASA declared itself insolvent on 10 February 2002. The company was only making partial interest payments (approximately 40 per cent of the interest on existing loans) on a *pari pasu* basis depending on the cash available.[16] In September 2003, AASA and its shareholders submitted a debt restructuring proposal to its main lenders. The latter initially refused, arguing that a restructuring proposal would only be open to consideration once the contract with Argentina had been renegotiated and the company's viability re-established. Nonetheless, a partial agreement on the restructuring of the company's debts was signed on 15 July 2014. It provided for a reduction in the bank debt, which therefore stood at US$655 million in September 2004. At the end of 2004, the stalling of negotiations between AASA and the Argentinian state prompted the International Development Bank and the SFI to propose their involvement in the negotiations. In March 2005, the Argentinian Economics Ministry agreed, but after several meetings in April and May 2005 the negotiations failed, which made a final restructuring of AASA's debt impossible.

As for the Argentinian state, in February 2002 the Economics Ministry had created a public service concession and privatization contracts renegotiation committee (CRCOSP), tasked with presenting the government with renegotiation proposals within 120 days. In April 2003, CRCOSP proposed a temporary six-month agreement which provided for a 10 per cent increase in tariffs to top up the fund used for extension works (*fideicommis*): in return, the company was to suspend the procedure initiated with the ICSID.[17] AASA rejected this agreement, which was put forward against a background of presidential elections.

On 25 May 2003, Nestor Kirchner became president. As Menem's opponent in the elections, he was very critical of the privatizations undertaken in the 1990s. On 16 July 2003, the Public Service Contracts Renegotiation Unit (UNIREN) replaced CRCOSP, under the dual supervision of the Ministry of the Economy and the new Federal Planning Ministry. In September 2003, the Renegotiation Unit received a report from ETOSS condemning the poor level of fulfilment of the contract; AASA nevertheless received a proposal for a temporary agreement from the Renegotiation Unit, including an ambitious investment programme with no tariff increases. Signed by the company on 11 May 2004, it was supposed to be a first stage, before a final agreement to be concluded within the year.

AASA was unable to convince either the Renegotiation Unit or the tutelary ministries of the need to increase tariffs or to downgrade the investment plan in order to re-establish the concession's financial equilibrium. Since the temporary agreement had come to an end and the other negotiations conducted in the first half of 2005 had gone nowhere, the AASA annual general meeting of 22 September 2005 called for the contract to be cancelled on the grounds of fault by the licensing authority. The government rejected this call and decided to consider a takeover of the company by local economic groups. This was a failure, since none of the potential buyers was able to obtain from the government the promise of a minimum tariff increase. Finally, on 21 March 2006, the Argentinian state announced the cancellation of the contract on the grounds of fault by the concession holder, claiming non-compliance with the threshold for nitrate content in the water distributed in certain sectors of the concession's southern zone (Decree 303/06). This measure was immediately followed by the creation of a new entity: Agua y Saneamientos Argentinos SA, which took charge of the services on the very same day (Decree 304/06).

In the course of this final period, economic conditions having become highly unfavourable, the operator sought to re-establish the contract's financial balance, but its efforts proved insufficient to avoid a breach of contract. The Argentinian government was increasingly overt in its role as an instigator of global social policy (subsidies for the *Jefes y Jefas de Hogar Desocupados* plan) combined with local applications of sectoral policy (*Agua + Trabajo* plan). This final period of crisis also coincided with a rise in the power of municipalities and user associations, whose role was considered critical to the viability of a new model for the management of water and sewerage services in Buenos Aires. While the economic crisis of late 2001 put an end to the peso–dollar parity and the economic balance of the contract, the complex negotiation – from 2002 to the end of 2005 – reveals as well a loss of legitimacy of the operator, deeply rooted in Argentina's political history. In order to understand why this process has ended in a cancellation, three intertwined levels have to be considered: the social and political environment, the initial characteristics of the contract and, finally, the behaviour of the company.

The gradual loss of legitimacy of the international operator

First are the changes in the social and political context. Since the 1940s, Argentinian society had been structured by Peronism, a populist national model notably characterized by a trend towards social homogeneity and political subordination to a leader (Svampa, 1999). In the second half of the 1980s, rising inflation brought this model into question. In 1991, price stability via the parity plan constituted one of the pillars of the consensus that the Menem government enjoyed (Minujin and Lopez, 1995), which enabled it to undertake large-scale privatization of public companies. Carefully prepared and drawing upon previous experiences, AASA's concession attracted no opposition. As the consensus around government action eroded with time and changing circumstances, social and political organizations – whether new or of "classic" peronist origin – gradually returned to centre stage (Merklen, 2008). This revival, with its oppositional and vindictive colourings, took several forms: "judicialization" of the contract (associated with the installation of collective water meters), demands by consumer associations (during the 1997 renegotiation), and opposition to the work schedules by certain municipalities in Greater Buenos Aires. After the 2001 crisis, the legitimacy of the foreign companies came to be contested by the new president Nestor Kirchner himself, who condemned the liberal policies of the previous decade and accused the major foreign private corporations of having "pillaged" the country (Akhmouch, 2009). This ideological stance made any real negotiation of the contract difficult, especially as the government refused to envisage a tariff increase that might exacerbate inflation (Zibechi, 2005).

The initial contractual conditions are the second aspect of the operator's loss of legitimacy that lay in the characteristics of the contract itself. Although the outcome of an extensive process of preparation by the Argentinian authorities, international experts and the company itself, certain "theoretical" foundations of the contract came into conflict with real-world practicalities. The contract was supposed to be comprehensive and guaranteed by independent regulation.

However, unforeseen factors quickly emerged and led to contractual changes. The incorporation of new environmental standards in 1994 led to a tariff increase that was in principle impossible in the first five years of the contract. Even more fundamentally, at the beginning of the concession, the opposition by the inhabitants of areas without a water supply to the economic terms of their connection (with the entire cost being borne by them) led inevitably to a renegotiation of the contract, since extension of the water systems was a major priority of the concession. What appeared in retrospect to be a necessary and effective adjustment to an initially poorly conceived clause became a source of criticism and helped undermine the image of the concession, leading to an uninterrupted sequence of negotiations between 1997 and the end of the contract. The implementation of the regulatory framework moreover required good definition and a logic of partnership with an efficient and competent regulatory body (ETOSS). However, there had been no regulator before the concession, and therefore no experience in the role: the original ETOSS staff consisted primarily of former employees of the public operator, with no experience in the principles of regulation. Moreover, the powers given to the regulator were not very extensive, since the contract was signed by the state and the consortium, which together settled the important questions. This situation was not conducive to the establishment of a collaborative approach. In the course of time as the concession evolved, the regulator gradually turned into a pernickety censor, which led to numerous hold-ups.

Third are contributions of the operator to its own loss of legitimacy. Despite undeniable results in the modernization of the technical system, in the extension of the water service to peripheral areas, in the financial management, as well as innovations in social engineering, AASA also contributed, in some of its choices or actions, to its own delegitimization. In order to guarantee the quality of the engineering studies at the start of the concession, AASA chose to sign a contract with one of its subsidiaries and a British engineering company, whereas familiarity with the terrain of the local design offices may well have avoided certain errors of judgement (such as the significant underestimate of consumption levels). In addition, this decision excluded local engineering firms from the market, potentially arousing their hostility. Likewise, the inclusion of Argentinian firms in the AASA would have seemed to be an essential step in preparing the tender bid. However, the two Argentinian companies that joined the consortium (SCP and Grupo Meller), from outside the water sector, were simply present as shareholders. When these firms sold off their holdings, Suez (ex-LDE) and AGBAR did not look for other local partners: the consortium thus found itself without local economic backers when problems arose.

By offering the lowest tariff, AASA acquired its initial legitimacy as operator of the concession. The contract stipulated that the tariff could not be increased during the first five years; yet a tariff hike took place less than a year after the start of the contract. Although this rise was justified by new requirements introduced by the government, and received government approval, it fed suspicion that the bid was opportunistic, a suspicion that grew with each of the subsequent tariff increases in the course of the concession.

In order to extend its services into disadvantaged areas, AASA, in collaboration with the regulator and the municipalities concerned, set up the MPG programme, which was based on the principle that provision of services could not be envisaged in these neighbourhoods without first resolving questions relating to urban planning (recognition of property rights with a public domain facilitating the layout of pipes under the streets). This model relied on the involvement of the municipality and the inhabitants. Its introduction and implementation bore witness to efforts to cooperate and to take account of local specificities. It also contributed to the genesis and development of expertise specific to the group in this domain. However, this programme also contributed to its promoter's loss of legitimacy, insofar as it showed that the development of services in poor neighbourhoods did not depend on the specific skills expected of a large international corporation (high technical expertise and capacity to raise substantial capital): the techniques could be directly implemented by teams of inhabitants following brief training. As a result, the programme's success had a paradoxical effect. It gave the federal state the opportunity to reassert its legitimacy in the sector with the subsequent development of the *Agua + Trabajo* programme, which it claimed to have devised, and in which AASA was now no more than a partner. Ultimately, this success showed that the issues relating to water and sanitation had changed, and that it was no longer necessarily essential to mobilize the capacities of a large international company, an assumption that had been one of the premises of the contracting-out process.

Conclusion

At the beginning of the concession, the private operator's legitimacy relied on a threefold capacity required by the tender: the capacity to implement expertise and technologies; the capacity to operate and improve services, and financing capacity in order to develop the network in areas without water and wastewater services. The concession was based on the principle of full cost recovery, intended to enable the company – through the tariff charged for its services – to repay the loans contracted in order to achieve the ambitious targets assigned to the concession throughout its entire term.

Over time, this model ran into several problems, in particular the problem of financing the service extensions. Very rapidly, from 1997 onward, the principle of financing all costs through tariffs alone, which lay at the heart of the concession, came under pressure. The operator's role then inevitably changed. While it remained the guarantor of the continuity of the existing services, its role in the extension projects was limited to its participation in development programmes in deprived peripheral areas of the concession, programmes in which it was initially the instigator and subsequently only a participant. Because of the results achieved by the company on the one hand, and the occurrence of the 2001 crisis on the other hand, the priority shifted towards the extension of the services to populations with few resources, which had fallen victim to the crisis. The three capacities (technical, managerial and financial) on which the private concession's initial legitimacy was founded were no longer necessarily required: a new operator profile was emerging, which not only differed from that of the large, specialized international enterprise, but one to which such an enterprise might not even be able to adapt while still meeting the requirements of its shareholders.

Appendix 3.1

Pricing and fee system

Prices

Bills were based on a bimonthly basic tax (TBB) depending on the category of user (residential, non-residential and unoccupied land) and on the characteristics of the property:

$$\text{TBB} = \text{MAX}\ \{\text{K}\star\text{Z}\star\text{TG}\star(\text{SC}\star\text{E}+\text{ST}\,/\,10),\ \text{TBB min}\}$$

K: modification coefficient (initial value: 0.731)
Z: zonal coefficient
TG: general tariff for each service
SC: total surface area
E: factor including the type of habitation and date of construction
ST: surface area of the plot
TG and SC differ from one category of user to another.
In addition: ETOSS tax (2.67%) and VAT (18%, then 21% from April 1995).

Connection and infrastructure charges

- The charge for connection to the service applied to any new connection (drinking water or sanitation) or for its renovation in sectors already supplied. At the start of the concession, the minimum amount was Ar$135 (water) and Ar$200 (wastewater).

The purpose of the infrastructure charge was to finance the cost of the network built by the concession holder in the extension sectors and of the corresponding new connections. It was only applicable to new users and was divided into equal bimonthly instalments spread over two years. At the start of the concession, the average benchmark values of this charge were Ar$325 (water) and Ar$460 (sanitation).

Notes

1 In September 2005, the shareholders of the company responsible for the contract, after years of negotiations, had called for a cancellation.
2 Following a period of military dictatorship which ran from 1976 to 1983, Argentina regained democracy with the election of Raul Alfonsin. The final year in office was marked by a profound hyperinflationary crisis which led him to leave office a few months before the end of its mandate to the winner of the next presidential election, Carlos Menem.
3 In 1993, OSN had more than 8,000 employees for a city population of 8.6 million; a supply rate of 70 per cent in potable water and 60 per cent in sewerage services across the metropolitan area as a whole (100 per cent in the city centre). A large proportion of the pipes had exceeded their depreciation period, industrial waste was not being treated and the customer data were obsolete.
4 In the rest of the chapter we will refer variously to Lyonnaise des Eaux or Lyonnaise; and AASA or the operator.
5 This setup for the regulatory body was supposed to ensure its independence; it was funded by a 2.7 per cent tax added to customer bills.
6 Created at the beginning of 1992, the Argentinian peso (Ar$) was aligned to the US dollar: from this point on, local currency was only created in direct line with dollar inputs. This system (called the currency board) was intended to strangle inflation and revive the economy. It favoured the national economy as long as the dollar was low, as it was in 1992, and as long as the deficit of the Argentinian state was under control and the local economy growing. However, following an initial alert with the Mexican crisis of 1995, the sharp rise in the dollar in 1998 put pressure on the system. After generous proceeds from the privatization of many state-owned enterprises the state faced again a deficit, leading to the crisis of 2001.
7 Litre per day, per head.
8 The People's Defender is an independent national institution established in the constitutional reform of 1994. It is responsible for the "defence and protection of the rights, guarantees and interests protected by the National Constitution and by the laws, against the doings, acts or omissions of the Administration, and for monitoring the exercise of the public administrative functions".
9 The appropriate return was assessed via an ATD (*Adecuada Taza de Descuento*) rated at 9 per cent (real value) and 11.5 per cent (nominal value).
10 The EFNQ is the result of the following calculation: $(C+I)-R$, where: (R) represents revenues, (I) planned investments and (C) the operational costs of the services and the concession holder's fair rate of return. At the end of the first five-year period of the contract (31 December 1998), the EFNQ was valued at Ar$823 million: this surplus of investment over revenues was taken into account in setting the terms of the second five-year period.
11 Grupo Meller and SCP sold their shares at three times the original purchase price. This sale was made for financial reasons. SCP was experiencing rising debt, since its revenue forecasts had been heavily affected by the local effects of the Mexican peso crisis of 1995/1996.
12 Between January 2001 and December 2005, cumulative inflation stood at 165.4 per cent for wholesale prices and 74.1 per cent for retail sales, and average wages rose by 58.4 per cent.
13 The *neighbourhood community* chose its representatives and submitted a request to the municipality for incorporation into the service. The *municipality* was responsible for opening up the streets. The *beneficiaries* provided labour for the construction, sometimes in exchange for the payment of Ar$150 provided via the municipality within the framework of the *Jefes y Jefas de Hogar Desocupados* social programme introduced in 2002 by the *Argentinian government*. *AASA* provided the

essential technical training for the works and the equipment needed for their implementation (including pipes). The *regulator* coordinated the actions of the community, the municipality and the company.

14 ENOHSA was the state agency authorized to contract multilateral loans and to manage the financial resources allocated to the sector.

15 The *Agua + Trabajo* programme entailed the creation of worker cooperatives which provided the labour to carry out extension works in disadvantaged areas. Each cooperative consisted of 16 workers, half of whom were beneficiaries of the *Jefes y Jefas de Hogar Desocupados* plan. *ENOHSA* provided the funding and developed the project with the *municipalities* and *AASA*.

16 All the creditors are treated equally.

17 The request to open a procedure was addressed to the ICSID (International Centre for Settlement of Investment Disputes) in April 2003 and was registered on 17 July 2003 under reference ARB/03/19.

4 Water in Paris

Polity and politics

Dominique Lorrain

Being part of a society entails the sharing of legal rights, values and experiences of varying significance (Nora, 2013; Rosanvallon, 2004). The points around which collective identity can begin to crystallize vary from one country to another. They may be secularism, national identity or the organization of public services, but polity and politicians are always present in the picture. The argument developed in this chapter is that even in activities with a long history, which may seem stable, nothing is ever frozen. Needs change, new technologies appear and politicians have to redefine their meaning.[1] It is therefore necessary for the factors that underpin community identity to be regularly renewed. When it comes to this primary matter of constructing meaning in collective action, urban technical systems raise an interesting problem. They represent long-term constructs that cut across the short-termism of contemporary societies; this is equally true of the directly material components – the infrastructures – as of the institutions. With time, these collective constructs become natural and the original political principles and choices fade, so that the history of urban networks could be cited as an illustration of the theory of path dependency (Thelen, 2003). In fact, these constructs are less technical and natural than might initially be thought. The political principles they embody need to be reactivated periodically in order to make action intelligible and legitimate.

The municipalization of water in Paris in 2010 offers a textbook instance of these general questions. Its story first of all illustrates the thesis of socio-technical systems that build up over time to form "models" with institutions and main actors (Lorrain, 2005). Around 1850, when it all began, water as a technical system was in its infancy. A new demand was emerging in the big cities – everything was started from scratch. Paris was a symbolic arena in which the encounter between state, city and the very new Compagnie Générale des Eaux would take place. This story can be broken down into three phases if we consider the legal framework. The first long phase, from 1860 to 1984, corresponds to the history of the invention – and then the stabilization – of a model.[2] This applies to the morphology of the technical system and to the lifestyles of Parisians with the arrival of household water: hygiene and progress (Goubert, 1986; Jacquot, 2002). It also applies to relations between the city and the operator, in which a contractual system was developed and bedded in. The second phase began in 1984, with the signature of a 25-year licence contract that extended the partnership. The French urban service model reached its apogee, with progress in the institutions and in the prerogatives of the operators; Lyonnaise des Eaux joined its main competitor to manage water in Paris. The third phase – the municipalization in 2010 – represents an absolute counterpoint to this 150-year history. How should it be understood? Was the move away from an urban service model in Paris that may have seemed stable a punishment for operator failure? Does it reflect a clear desire among Paris politicians to adopt a different way of organizing urban services? Or is it perhaps an illustration of the permanent political dimension of these activities? In this case is it the outcome of shifting political alliances or should it be seen as the reflection of a deeper need to overhaul the founding principles?

Revisiting models, deconstructing institutions

When it comes to water and sewerage services, France is counterintuitive. This country, where the state is a dominant presence, even in market activities, developed a model for the management of urban infrastructures under local authority oversight, assisted by private companies which in time grew to become world leaders. This solution emerged over time, discreetly, without big debates in Parliament, ultimately to become an original construct with its own coherent character, and a form that can be modelled: an urban service model (Lorrain, 1992, 2005; Stoffaës, 1995). At the beginning of the 1990s, it may have seemed that this was a stable solution that could develop, as evidenced by its reproduction in many French cities, its spread to new urban services and above all its international expansion.

An urban service model may be defined as a simplified representation of a number of phenomena which seem, at first sight, to possess no consistent characteristics, but which take on order in the light of long-term observation and through the use of general categories that encompass the particularities of each micro-object. Following initial research on urban services in France, followed by a comparison with Great Britain and Germany, we developed an interpretation based on three levels: institutional architecture, principles of action and political culture. From this, we formulated three major descriptors: institutions and culture and, between them, techniques of action (contracts, funding methods, etc.). This result, derived from the study of the object without reference to any theoretical current, proved to be fairly congruent with the works of institutional economists. In a seminal work, Douglas North provides an analytical framework to appreciate the role of institutions in economic performance. He makes a difference between formal and informal institutions (North, 1990: 36 and 46). The former lay down property rights: "they are the rules of the game in a society.... In consequence they structure incentives in human exchange" (North, 1990: 3); they result from choices and compromises between the different stakeholders and, for this reason, do not change overnight. Informal institutions – or values and culture – are long-term accretions, shared by the stakeholders and generated by education and experience, and therefore are also slow to change.

However, this analytical framework raises two problems. First, the notion of formal institution is a very broad category, which includes the organizing principles of large institutions and the rules of organizations; it encompasses international treaties, constitutions and large-scale national laws on the one hand, and a lot of rules and instruments on the other hand, whose number and influence is growing. It is true that these institutions have in common their formal character, which distinguishes them from informal institutions, but otherwise they share neither the same order of generality nor the same development processes. Second, this theory emphasizes the long-term and incremental process of institution building, which reinforces the thesis of path dependency; yet the reason why institutions survive is because they adapt. This analysis, combined with our initial three-level representation, prompted us to deconstruct the category of formal institutions and to introduce second-rank institutions: decrees and circulars that complement the laws, technical standards and certifications, calculation formulae, accounting principles, management and financial ratios, and ethics charts (Lorrain, 2008a).[3] These may be defined as institutions that are more familiar to the stakeholders, structures dedicated to problem-solving and action. The first advantage of this deconstruction is to distinguish between different categories. Above all, it provides an explanation for the problem of change – and the other side of the coin, namely path dependency. Second-rank institutions are the primary locus of innovation; they are familiar to the actors and action orientated; the stakeholders constantly build "small" institutions in order to solve practical problems. The adjective "small" also reveals another property: being less sanctified than the "big" institutions, they are easier to reform. This deconstructed interpretation enriches the initial description by differentiating three categories of institution: formal institutions or framing institutions, informal institutions or silent institutions (revealed by crises), and second-rank institutions or active institutions.

What can be said about the French urban service model through the prism of this interpretative framework? The particular arrangement devised for water in the second half of the nineteenth century spread to other network services, with the result that we find the same institutions and actors involved elsewhere. Municipalities are in control (i.e. city mayors) in the French political system. The operators are large private firms that contributed to the introduction of certain market principles into the management of essential goods, principles that were adapted and tempered to make them politically acceptable. The state contributed to the development of the rules, funding arrangements and monitoring methods to maintain overall balance. Thus the system moved from the original plain concessions to more balanced contracts between public authorities and private operators (leases, cost-plus schemes, operating contracts). In short, it moved from concession practices to public services contracting.[4] It is an effective solution, suited to the French situation, with its numerous municipalities in many cases lacking critical mass. The major operators provide the resources they lack, and by spreading the "right" solutions from one territory to another they bring efficiencies through economies of scale. However, this structure has one weakness: a lack of political adherence. The broad principles were set out in the second half of the nineteenth century with no real debate, after which events unfolded from the initial conditions. The model, adapted through a proliferation of second-rank institutions – contracts, funding methods, selection rules, information formats – was designed to resolve problems or neutralize criticisms (as in the inter-war period). If asked "How is water governed?", one could reply: pretty much automatically. The main agents of this story were a few top politicians, senior civil servants and the company bosses, whereas citizens/users remained on the edges of the system; they expressed themselves tacitly by accepting the choices of their political leaders. However, a question remains: Are these "small" institutions and the informal principles of the model still understood?

In certain respects, this system is out of line with the deep values of French society – the state as the vehicle of the public interest, criticism of the profit motive, mistrust of big private companies. France is built on large-scale laws and codes, whereas public service contracting belongs to a different tradition – the spirit of the common law. If some kind of canon were to eventually emerge, it may be seen in retrospect as the outcome of legal decisions – rulings of the "*Conseil d'État*".[5] This incongruity is all the greater in that the model was constructed discreetly. No doubt discretion was necessary for the companies to be able to develop, but the other side of the coin is that there is no corpus justifying the choices, their originality relative to other models and the potential advantages. In evidence of this absence, the mayors responsible for signing and renewing contracts provide no public explanations of their grounds for awarding them. All they say is, "I made a good deal (on the price) and protected the interests of my voters-users". Similarly, the companies say little about the general framework of public service contracting; they simply provide solutions to problems. This may be seen as the attitude of pragmatic engineers applying an implicit division of labour – it's up to the politicians to explain their choices; the companies are just there to solve problems. This explains the paradox of having companies that have been part of the landscape since 1850 to 1880, and yet they regularly face questions on a primary issue of legitimacy. The grammar of state public action and the spontaneous "big-is-bad" culture combine to override other representations. There is no theory of public service contracting in France. That is the paradox of the French urban service model. It has developed without updating its principles and constitutive values. This can result in misunderstandings, as evidenced by the example of Paris.

Reinforcement of a model: public service contracts (1985 to 2009)

The organization of the water industry in Paris followed the scheme introduced in the late nineteenth century, with potable water production and management of the distribution network carried out in-house by a municipal department, and the support of Compagnie Générale des Eaux on the billing side, replacing meters and installing connections.[6] Rainwater and wastewater collection within the city area was managed by another municipal department, subsequently supported by a second-tier public entity, namely SIAAP (Bellanger, 2010).

This arrangement was modified in 1984 by the Mayor of Paris Jacques Chirac, who had been elected in March 1977. The political background here is of some importance. The election of the Mayor of Paris was a highly symbolic event, since it marked the end of the capital's administration by central government-appointed prefects. It gave the mayor political influence as boss of France's leading city, a world-ranking metropolis. Jacques Chirac used this position to serve his national ambitions and his comeback strategy following the election of socialist president François Mitterrand in 1981. The left undertook a large-scale nationalization programme at the very moment when the liberal cycle instigated by Great Britain was just getting underway. The French right, briefly disturbed by the loss of a power it had held unshared since 1958, drew inspiration from the liberal winds blowing across the Channel and directed it to the cities that came under its control following the 1983 municipal elections. The new strategy, "reconquer France city by city", was to demonstrate that there are other ways to manage than direct administration. The right was keen to emphasize its difference from the nationalizing left. Several mayors newly elected in 1983 contracted out network services previously managed by municipal boards: Michel Noir in Lyon, Alain Carignon in Grenoble, René Bousquet in Nîmes, as did a few others in big cities like Sète, Orleans, Amiens, Toulon, Caen, etc. Paris became a political powerbase for its mayor. It brought him resources and maintained his international profile. His organizational choices were held up as an example of what he could do for the country as a whole if the electorate gave him its votes in future elections. The city contracted out its car parks, its funeral services, its public lighting, its drinking water, part of its waste collection and so on.

For potable water, the contracts were awarded to three companies. Production was taken over by Sagep, a semi-public company owned 70 per cent by the municipality, 14 per cent by each of the distributors and 2 per cent by para-municipal firms. Distribution (i.e. network management, billing and customer relations) was contracted out to the two major French water companies, with a division on geographical lines, the right bank going to Compagnie Générale des Eaux (via Compagnie des Eaux de Paris) and the left bank to Lyonnaise des Eaux (via la Parisienne des Eaux[7]). These contracts were awarded without competitive tendering and any subsequent independent system of regulation. Sagep was responsible for monitoring, a role which it performed with little stringency. In fact, the meaning of the term "monitoring" is questionable in circumstances where the monitoring agency is also a direct participant, and the two companies it is supposed to monitor are among its shareholders. It was only a national decree of 1995 that laid a requirement on contract holders to provide the mayor with an annual report (Cour des Comptes, 2003). In the mid-1980s, the model continued to be structured by its constitutive values – the importance of *intuitu personae*, trust, self-regulation. The theory of the agency, with its formal distinction between the commissioning principal and executive agent, was unknown in France. These practices, which may seem surprising to outside observers from other traditions, were more broadly consonant with France's politico-administrative system of the time. The state-owned companies were run by senior civil servants with political and managerial skills, who were considered to represent the public interest and were therefore largely spared formal debate and monitoring procedures (Lévy-Leboyer and Casanova, 1991; Kuisel, 1984). The system was self-regulating in a relatively closed circle, without a hint of concepts such as independent regulation and public debate (Vickers and Yarrow, 1989; Baldwin *et al.*, 1998). At the end of the chain, the informality of local operations merely reflected these more general principles of French economic management. Nonetheless, this system produced results. For example, the deputy mayor in charge of the sector and chairwoman of Eau de Paris acknowledged that "technically it wasn't too bad (in 2001); system efficiency was not very high but that was from a starting point of a 22% loss rate at the beginning of the contract".[8]

Some believed that the objective of the two private companies was to build up their holdings in the semi-public production company and to take over the whole potable water cycle. This is not what happened. First, Sagep executives, who were drawn mostly from the Municipal Department, began to hanker after independence. Second, the mid-1990s marked a moment of transition after a decade of sharp expansion in private sector operation; there were criticisms of the water companies, rules were established and even politicians favourable to the private sector found themselves with less freedom of choice.[9] Finally, Jacques Chirac's strategy for national takeover from his Paris powerbase succeeded in 1995, with his election as President of the Republic. He was succeeded by a mayor from the same political party, but times had changed; there was now greater oversight of the management of the city than in the Chirac years, and the new mayor quickly saw his power fade.

At the next municipal elections in 2001, the mayor was beaten by Bertrand Delanoë at the head of a coalition of socialists, communists, ecologists and others. From the start of his term, it was decided to "take public management in hand, to restore public control and re-establish prerogatives".[10] Inspections were carried out in a variety of public services. In the same period, *la Cour des comptes*[11] prepared a special report on the management of water and sewerage services in France, which made several comments about the potable water service in Paris. The city began negotiations with the contract holders in response to the comments of the Court, on issues such as reducing production volumes in the light of falling consumption, speeding up a schedule of works on the network to reduce losses and improving the information provided in the annual report to the mayor. With regard to system losses, the operators had not fully used the provision for renewal work allocated in the cost calculations. Moreover, since the distribution pipes were installed in the sewers – a legacy of the large-scale construction works carried out by Haussmann and Belgrand – the lost clear water was not returned to the natural environment, but was carried by the emissaries through to the treatment plants. In other words, the negative impact of the water losses was twofold: more upstream water to be treated to become potable, and more clear water in the downstream treatment plants. Following negotiations between the city and the operators, amendments were added to the contracts: the companies undertook to carry out a loss reduction programme and to replace 93,000 water meters, funded from the provision surplus.

These undertakings were fulfilled. Over the period of the contract more than 1,100 km of pipework were renovated or repaired, slightly more than half the total length, rejuvenating the infrastructure. The efficiency of the distribution network improved, from 76 per cent in 1985 to 96.5 per cent in 2009, making Paris one of the world's most efficient cities in this respect.[12] This programme saved around 617 million cubic metres of water, the equivalent of three years' annual consumption.[13] To achieve this result, the operators installed an info-system of sensors to track the technical functioning of the system (Curien, 2000): the 94,000 water meters were fitted with a remote reading system and a database was set up for the 2,000 km of pipelines, providing information on the condition of each section, the nature of the work required, follow-up plans, etc. Finally, in application of a public health regulation, all the connectors made of lead – a total of 66,000 units – were replaced. This programme was completed at the end of 2009.

The price of potable water, a sensitive subject characterized by a degree of misinformation,[14] rose from an index of 100 to an index of 276 (i.e. an increase of 176 per cent). Allowing for inflation (+70 per cent over the period), the constant price doubled over 25 years (see Table 4.1). However,

Table 4.1 Prices over the period 1985 to 2009

	1985	*2009*	
Inflation	100	170	
Potable water (current price)	100	276	increase 176%, multiple 2.76
Potable water (constant price)	100	206	increase 100%, doubling
Global bill (current)	100	390	increase 290%, multiple 3.9
Global bill[a]	€0.74/m^3	€2.89/m^3	multiple 3.9

Source: Barraqué (2012: 908).

Note

a J.L. Trancart, note sur l'Eau à Paris, working paper (6 February 2012, p. 7).

this "potable water" price represents only part (40 per cent) of the total bill paid by users, alongside wastewater (31 per cent), and fees and taxes (29 per cent) (see Appendix 4.1). The overall bill had gone up 3.9 times in current prices owing to increases in the costs for wastewater (+6.7 per cent per year) and fees and taxes (+8.4 per cent per year), mainly reflecting the efforts going into decontamination measures. The rate of increase for potable water was less, at 4.5 per cent per year.

Municipalization: 2010

"Paris water is of excellent quality, it enjoys security of supply from several sources, its networks are well maintained and its price remains below the national average", said the Mayor of Paris in 2007.[15] Five years later, a report written by the City of Paris declared that: "The City of Paris has taken full public control of the water service in order to provide management that is *rigorous, transparent* and *efficient* on behalf of the men and women of Paris" (Mairie de Paris, 2012: 14). If words mean anything, this implies that for the writer of this piece the management had previously lacked those qualities of rigour, transparency and efficiency.

> Today we see that (the choice of the method of operation of public services) is not subject to the same (symmetrical) formal obligations from one method to another. Thus, the choice of direct management does not require prior justification, whereas this comparison is compulsory in other countries.[16]

In the run-up to the 2008 municipal elections, the outgoing mayor announced that, if elected, he would bring water under municipal control. There was no reaction to the announcement, more a sign of indifference than agreement. The two operating companies, reassured by their positive results, believed it was primarily a political tactic which would not be implemented.[17] The mayor was re-elected in the first round, and took advantage of a window of opportunity to do what he had announced: his re-election in March 2008 gave him legitimacy and the contracts were due to expire in December 2009. A new public entity was set up to manage combined potable water production and distribution and rainwater transport: Eau de Paris. The transition from private to public took place quickly and smoothly.

Efficiency measures

We will begin by examining the industrial case, on the assumption that when people overturn an existing organization it is in order to do better, especially as the primary principle is still to serve the public interest. At the time of municipalization, it was announced that it might be possible to save €76 million a year by grouping the three contracting companies, leading to an 8 per cent price cut for customers. According to the published data, this goal has not been achieved (Appendix 4.2). The price of potable water was cut on 1 July 2011, but since it was part of a total bill in which the other components increased, the final bill was higher (Table 4.2). On 1 January 2013, the average bill was 7.2 per cent higher than in 2009. In any case, the impact of a cut in the cost of drinking water upon household budgets could only be small: the 8 per cent reduction corresponded to a saving of €0.0837 per cubic metre consumed[18] (i.e. €7.50 per household per year).[19] In addition, once the cost of water was considered in the context of the cost of essential goods in Paris, the promised reduction had all the features of political hype. In the housing sector, the price per square metre of old residential property – the largest cost item in household budgets – more than doubled between 2001 and 2010. Long-term statistics show that between 1965 and 2000, the national price index for old residential property grew at the same rate as "disposable household income". These two indexes parted company in the 2000s.[20] Paris is an extreme example. In 2010, the average price of old residential property was €7,500 per square metre[21] as compared with €3,059 per square metre in 2001.[22] This decade emerges as one of an explosion in real estate costs, with all the social consequences that entails: indebtedness, gentrification, etc.[23] When the average price increases from €3,000 to €7,500 per square metre, the same $100\,m^2$ apartment costs €450,000 more! That is a totally different order of magnitude compared with the cost of water. This reference to housing can be extended, if we consider the fiscal side. In 2008, without any debate, the City of Paris introduced a new departmental land tax of around €140 for an average three-room apartment. Coming on top of the existing land tax (*taxe foncière*) of €740, this represented an increase of 20 per cent. This sudden new tax was explained by the need to offset the collapse in tax revenues from property transactions which, after strong increases with the explosion of property prices in the 2000s, suffered the fallout from the subprime crisis. This departmental tax would be increased for a second time in 2011 (around €100 for the same average three-room apartment), accounting for 25 per cent of a global tax of €1,000.[24] We can see here the limitations of political rhetoric that focuses on one sector, when the real problems lie elsewhere.

Table 4.2 The total price of water in Paris

	2009	*2010*	*2011*	*2012*
Euros per m^3 for $120\,m^3$ per year[a]	3.07	3.10	3.18	3.29

Sources: Mairie de Paris, Annual reports on the price and quality of the public potable and wastewater services (2009: 16; 2012: 27, 28).

Note

a International scale comparing consumption by an average household of 2.4 people.

Regardless of these limitations, what factors might contribute to a cost reduction of €76 million per year, as it was announced? What are the margins to improve a technical system? First, the non-payment of dividends to the private operators saved around €10 million (on the basis of a net pre-tax profit of €13 million) (see Appendix 4.3). Second, greater efficiency in maintenance work through competitive tendering could bring savings of 10 per cent, but we are still a long way short, since the average investment was €60 million per year (Appendix 4.4a). Third, although organizational unification can potentially generate significant savings, these did not materialise. As noted in the 2010 and 2012 reports, the amalgamation of three separate organizations (the production company, the two distributors) was not something that could be done overnight: it consumes time and resources.[25] There are transaction costs.

It would seem that a large proportion of the savings came from administrative engineering. With municipalization, Eau de Paris could reduce its costs because it benefited from more favourable rules than those applied to the private operators: business tax, pension scheme, VAT repayment on construction work, corporation tax and so on (Babuziaux and de Leyssac, 2005). Moreover, Eau de Paris was no longer bound by a specific schedule of works under a visible and binding contract. A "performance contract" was approved by the Paris Council in 2009,[26] but in the absence of any cost accounting it is impossible to track its application; as a result, the board could defer investment and generate surpluses (Appendix 4.4a).[27] Paradoxically, we find the same kind of creative financial engineering for actors on the opposite side of the political spectrum: private equity funds. The results are more to do with playing with the rules than with industrial policies designed to improve performance in the technical system. If we look for a purely economic explanation for this municipalization, we find it in cash flow management. Drinking water distribution generates revenues of around €210 to €220 million a year, sums that could be collected by the Treasurer of the City of Paris and managed in a consolidated way. The separate financial statement for Eau de Paris is only produced at the end of the financial year. In the meantime, the city can use these fungible resources, and thereby reduce its reliance on short-term loans.

Transparency . . . what the financial statements say (and don't say)

There are three main components to a household water bill: potable water (40 per cent), wastewater (31 per cent), and taxes and fees (29 per cent). The last two components have gone up, but they are not managed by the water operator.[28] In other words, the water operator receives monies on behalf of third parties and transfers them. As a result, its turnover – as measured from water bills – is artificially inflated. We tried to use the financial statements of Eau de Paris and Paris municipality's specific water budget in order to deconstruct the water account and obtain a real picture of the activity of Eau de

Paris, with a breakdown of income and expenditure. This exploration of documents in the public domain may also be interpreted as a test to see what, in practice, the stated goal of transparency really means.

We began by examining the Mayor of Paris' specific water and wastewater budgets, on the assumption that the overarching authority, responsible for all the entities involved in the water cycle, would give us a consistent picture. Not at all. The city's water budget provides a wealth of detail exclusively on the items relating to its exchanges with its satellites.[29] We then went on to examine the three annual reports of Eau de Paris (2010–2012). These documents give us the figure for total revenues (based on bills paid and various services); once expenditure is deducted, we obtain the EBITDA,[30] and with depreciation and other exceptional operational items we obtain the figure for net income (see Appendix 4.4a). However, we are foiled by the opacity of the accounting system, because the proceeds include amounts collected on behalf of third parties, which are simply revolving transactions. Let us look at the 2010 statements (Appendix 4.4a). First, unadjusted reading: total sales €359 million, total expenditure €266 million, EBITDA €93 million (i.e. a ratio of EBITDA to revenue of 25.9 per cent). Since the organization had no major investment programmes, it could finance almost everything out of earnings.[31] Second reading: revenues of €358 million, minus deductions for third parties of €137.5 million, which impact upon revenues and expenditure. Real revenues fall to €221 million, expenditure to 129 million. EBITDA emerges at €92 million (i.e. a ratio of 41.2 per cent) (see Appendix 4.4b).

Once transfers are eliminated, we obtain two results: Eau de Paris has revenues of between €220 million and €210 million. At between 41 per cent and 46 per cent, EBITDA (gross operating profit) compares favourably with the high profits of UK water companies in the early 1990s. However, this picture of the figures is out of step in two respects. First, potable water represents 60 per cent of the total water sector turnover, yet potable water only accounts for 40 per cent of the bill. The figures don't match. We looked for an answer in SIAPP's annual report, but it provides no financial data. Second, the questions continue if we consider that the city sells around 180 million cubic metres of potable water (Appendix 4.4b); with a price of €3.29 per cubic metre, total revenues should be around €590 million, a long way from the €358 million before the deduction of fees and taxes. In short, the aggregates do not match up. We are missing a piece of the puzzle.[32] This totally public water system, under the guidance of a single player, is not transparent in the essential data: revenues, net earnings, investments, etc.

Although imperfect, however, these data may be used to question the policies pursued or omitted. The unconsolidated statement shows EBITDA of more than 40 per cent. What is this surplus for? The operator could borrow for investment and dramatically reduce its prices. Another possibility would be to pool costs and prices between the city centre and the suburbs, as happens in provincial cities. Paris is in an exceptional position, which explains its fairly low costs – an old system that was paid off long ago, pipes laid in accessible sewers which facilitate maintenance operations and explain the network's excellent output (achieved before it came under municipal control), a supply half of which originates in spring water gravity fed via aqueducts, which reduces the transport costs for this bulk resource,[33] very high housing density, which limits the length of the distribution network (2,000 km), and a small number of collective meters: 94,000 to serve 2.27 million residents, or three million including daytime users.[34] A wealthy city, Paris reaps the benefit of all these economic factors and can sell its water for around €3.12 per cubic metre,[35] whereas the price for the large federation in the neighbouring suburbs is around €4.80 per cubic metre, after renegotiation of the contract (Defeuilley, 2014). This difference cannot simply be explained by the operator's profit margins. Until recently, the same private operators worked in Paris and the suburbs. Municipalization has separated Paris from its suburbs, with each managing its "own" water. Paris goes a long way to fetch it, crossing suburbs without having to pay for rights of way. In the economic mechanisms of the day, we find survivals of the early nineteenth-century legacy, with an all-powerful capital that dominates its nearby territories and benefits from windfall effects. These very concrete realities, quietly embedded in economic laws that benefit the strongest, very largely eclipse the rhetoric about transparency and citizen power.

In other words, the system is still opaque. The city does not consolidate.[36] Three public institutions are responsible for water in Paris: the municipality, Eau de Paris and SIAAP. Each applies public accounting standards, but there are no bridges between these institutions and their accounts. It is up to users to do the work but we doubt that they will ever be able to produce a synthesis of all of this data. This would seem to be further proof of the importance of second-rank institutions and of the autonomy they can achieve as a kind of "automatic pilot". In the case of Paris, the person mainly responsible for this water policy, from 2008 to 2014 has pursued a genuine plan for political democracy in this sector.[37] She has dedicated time, introduced a number of resources – an information platform, online documents, a water survey – and yet the moment we try to describe this system of water governance using basic data, the difficulties begin. It has much to do with transparency, but the data formats are opaque. This is because the public accounts have their own categories: no consolidation, no cost accounting, no asset account (or balance sheet) that reflects the assets and debts of each organization. The accounts are based on budget years. The procedures for closing the accounts of programmes that run over several years are not fully standardized. It is impossible for the user to obtain a clear picture of investment policies. It is up to the reader – confused by an annual variation or by a difference between two figures that should be the same – to find an explanation. What might be acceptable in a community of specialists is inadequate to the political goal of placing the user at the heart of the system. The average user is not an expert on water. Users are likely to have the same expectations regarding policies for transport, parking, housing, air quality and bank charges, so if citizens are to be involved, they should have access to robust documents that give a clear picture of matters. Otherwise, these citizen participation procedures attract only people with a grievance or those who assume that public management is inherently desirable, whatever the situation.

The politician's simulacra

Ultimately, therefore, what we see is that price increases when it should fall; the much-touted transparency does not apply to essential economic questions. In short, there is a gap between the rhetoric that preceded municipalization and the policies that followed. This illustrates two major properties of the water sector, one economic in character, the other political; in fact they apply beyond the case of Paris and the water sector.

In a highly capitalistic industry where prices are not free, there is limited room for manoeuvre. This means that major initial investment is required before the activity can get underway. This economic property structures the sector over a fairly long period, and this limits the freedom of action of any operator, public or private. In addition, since water is an essential good, it must be universally affordable, and this has an impact upon pricing policy. Prices must be reasonable and not vary. That is another constraint. There are other activities which are also highly capitalistic – maritime transport, chemicals – but in these cases the prices are market-based and free; investors commit large amounts of capital but make significant profits; they have a certain freedom of action. The operator's leeway is further reduced when consumption volumes fall and impact upon revenues, because the fixed component of the pricing formula is small. This raises a problem of the sharing of fixed costs, which further diminishes the likelihood of a reduction in the price per cubic metre. In Paris, after increasing regularly for decades,[38] the consumption of drinking water has been falling steadily. Between 1960 and 1990, the volumes billed increased by more than 43 million cubic metres,[39] and then declined during the subsequent 20 years, despite the significant contribution of daytime visitors (Appendix 4.5).

The rational arguments that may explain the municipalization of water in Paris – the quest for greater efficiency or lower prices, poor provision by the private operators – do not hold up under scrutiny. Nor was it a response to pressure from citizens, as it was in Berlin.[40] So why the decision to change operator in order to achieve an equivalent or even poorer level of service? To understand the decision we need to look at political factors, factors that are too often neglected insofar as the continuity of the renewals gave them an air of naturalness. Even in a highly technical sector, where the use of industrial firms makes sense, even when the rules of the game seem stable, the political dimension is always present.

It is present first in the form of the political game: competition between people, elections, conflicts/alliances, victories/defeats. During the terms of office of Jacques Chirac (1977–1995) and Jean Tiberi (1995–2001), the principle of public service contracts was not questioned; the operating companies enjoyed unfettered freedom. This changed with the election of Bertrand Delanoë in 2001, at the head of a list including socialists, communists and ecologists. There were criticisms of the private companies, but no challenge to the contracting-out principle. In autumn 2007, in the run-up to the municipal elections, interchanges between the water companies and sources close to the mayor seemed to establish that the contract terminating in 2009 would not be challenged: "Don't worry, the contracts will be renewed, though of course with new terms of reference".[41] Things changed a few weeks later. The Mayor of Paris announced that, if re-elected, he would transfer management of the drinking water sector to a public entity (*Le Monde*, 7 November 2007, p. 9). What factors explain the shift? Serious misconduct by an operator? The emergence of a new technology that would change conditions? None of that. In fact the explanation is political, and has its origin somewhere between personal ambition and political alliances. The Mayor of Paris was preparing for the 2008 municipal elections – as was known – and more unofficially was considering his candidacy for the primaries in the 2012 presidential election, so he was positioning himself for his party conference. In Paris, the Ecologists and Communist Party officials were pushing for a rethink of the drinking water distribution contracts. In his political strategy, the mayor's first imperative was to keep his left wing happy. In other words, the first political dimension refers clearly to politics: alliances between parties to win an election and distribute positions of power. The undertaking to municipalize the potable water service satisfied the left in the outgoing mayor's coalition and helped with his alliance building. In this game of political barter, the incumbent's objective is to win a majority in the first round of elections to avoid a second round with a classic right–left confrontation, in which the result is always unpredictable. However, this explanation covers only part of the political dimension. Why did the bartering take place over water – a technical subject of only secondary interest to Parisians in their day-to-day lives – and not over transport, housing or the plan for a sustainable city?

This is where we come to the need politicians have to demonstrate that they are in a position to make choices and to have a direct impact on the real world. This need is particularly strong in France, since in the sphere of public action the country is experiencing a tragic transition. It was built around a strong and interventionist state (Rosanvallon, 1989), whose prerogatives have been steadily diminishing since the 1980s: decentralization in favour of local authorities, transfer of legal and financial prerogatives to Europe, the return of numerous state-owned enterprises to the private sector, and management based on market principles. In this new context, politicians govern less directly; they intervene by adjusting the rules of the game, by providing incentives. This new expression of power is out of step with dominant representations; it is less visible and its time frame is the medium term. However, the language of politics, in particular during electoral campaigns, continues to operate on the old register – resolution, decision, result – but now in relation to ever-shrinking spheres. In these circumstances, water stands as a low-cost political marker;[42] it implies an equation: public good = public management. Its management remains a symbolic issue, in which divisions between left and right can still be mobilized. This option is particularly attractive in that contracting out was not founded in France on a set of principles that justifies the reasons for employing private companies to manage essential goods. Despite the long history of public service contracts, the sales pitch available to politicians was weak, by contrast with Germany or the United Kingdom. In other words, the inadequacy of attempts to legitimize the contracting out of public services in France opened up an area of uncertainty which politicians could exploit.

Conclusion

Networked public services have a dual property. They are technical objects subject to physical laws, operating according to industrial and management principles. Senior civil servants and company officials tend to emphasize this aspect, particularly in countries where the rules of the game seemed stable. At the same time, networked services depend on polity and politics. Polity is understood as the higher sphere in which shared collective rules are set out. This "shared" condition is crucial, because it is what makes the rules acceptable and public action legitimate. In these conditions, public services contracting cannot be justified on purely economic grounds. It is not enough to do the job efficiently; the principle of the job must itself be legitimate, understood and accepted. The challenges to several public service contracts in Latin America may be interpreted as the expression of a breakaway from certain general principles underpinning the political societies in question (Merklen, 2008).[43] The transition from an economic argument to identification with a political project requires a process of public explanation, the essence of which may be understood by all, even in the case of quite technical subjects (Callon *et al.*, 2001). These processes are fairly poorly developed in France. The example of Paris provides a caricature of this need to construct a wider political legitimacy. The French urban service model has historically been founded on a special relationship between the mayor and the public service provider, with citizens involved only by default, through mediation by politicians and in the secondary figure of the user. For this reason, politicians occupy a central position and possess freedom of choice, and it is because the politicians are free that politics can intrude. In order to understand here how models adapt or bifurcate, we need to look at the circumstances, at the opportunities and at the balance of power; François Jullien speaks of the "potential" of a situation (Jullien, 1998). The opponents to public services contracting, in the case of Paris, took advantage of a window of opportunity: municipal elections in 2008, Socialist Party primaries in 2009/2010 and termination of the contracts in 2009. This freedom reinforces the tendency of politicians to focus on what is easy, symbolic and visible, even if the final result is no better than the previous solution.

Appendix 4.1 Data on the years 1985 to 2009 (public service contract)

Breakdown of average water price including tax:

Potable water	40% (production 19%, distribution 15%, municipal share 1%, meter rental 5%)
Wastewater	31%
Taxes and fees	29%

Average annual bill increase:

Potable water (production and distribution)	+4,5%
Wastewater	+6,7%
Taxes and fees	+8,4%

Sources: Bipe (2008), presented in Aquae, no. 46, November 2009, p. 4; Mairie de Paris, Annual reports on the price and quality of the public services for potable water and wastewater (2009: 14).

Appendix 4.2 Water bill in euro for 120 cubic metres: municipalization 2010

	2009	*2010*	*2011*	*2012*
Production and distribution	125.16	125.57		
Meter rental	20.08	19.72		
Total potable water	145.24	145.29		
Total bill[a]	368.93	372.17		
Price for 120 m3 per year[b]	€3.07/m^3	€3.10/m^3	€3.18/m^3	€3.29/m^3

Source: Mairie de Paris, Annual reports on the price and quality of the public potable and wastewater services (2009: 16; 2012: 27–28).

Notes

a On average, the total bill breaks down as follows: potable water (40 per cent), wastewater (31 per cent), taxes and fees (29 per cent).

b The price for 90 m^3 (average consumption volume in Paris) is slightly lower: €3.03 in January 2012 and €3.12 in January 2013. Mairie de Paris, 2012 report (op. cit.: 28).

Appendix 4.3 Economic data at the end of the public service contract

Revenues of the two distribution companies in 2009	€120 million
Pre-tax profit of the two distribution companies	€13 million
Revenues of the production company	€100 million

Source: Barraqué (2012: 908, n. 7).

Appendix 4.4a Operating account

	2010	*2011*	*2012*
Revenues (including fixed asset production)	358.9[a]	346.9	339.2
Expenditure	266.1	252.0	252.5
Including fees and taxes	147.8	135.9	127.1
Gross operating profit (EBITDA)	92.8	94.9	86.7
EBITDA/revenues	25.9%	27.3%	25.5%
Net earnings	76.0	75.2	62.8
Investments completed	57.0	35.4	76.0
Investments scheduled	(65.7)	(49)	(92.6)

Note

a Eau de Paris, Annual Report (2010: 61, "operations section"). We used this source because it gives expenditure, EBITDA and net earnings consistently from one year to the next.

However, the figures that follow in each report are not necessarily consistent. For example, for the 2010 revenues we see (p. 62) that "revenues stood at €361 million, of which fixed asset production accounted for €2.9 million.... Real operating revenues were €358.1 million". True, the differences are small, but we find ourselves with three figures for activity! For the taxes, duties and associated payments item, the differences are larger. In the operations section, the figure is €147.8 million. On the next page (p. 63), which gives a breakdown of operating revenues, the amount shown for fees is €137.48 million.

Appendix 4.4b Account recalculated removing third-party revenues (wastewater fees and taxes)

In € millions	*2010*	*2011*	*2012*
Total revenues (millions €)[b]	358.1	352.2	345.1
Including fees and taxes	137.5	134.6	136.5
Water revenues (after transfers)[c]	220.6	217.6	208.6
Expenditure (after transfers)[d]	128.6	117.4	116
EBITDA	92	100	93
EBITDA/revenues	41.7%	45.9%	44.6%
In millions of m^3			
Volumes registered/billed to subscribers[e]	184.1	179.0	177.0

Notes

b Eau de Paris, Annual Report (2010: 62, "operating revenues"); Annual Report (2012: 78–79).

c Calculated by subtracting "fees and taxes" from revenues.

d Calculated by subtracting "fees and taxes" from expenditure, as identified in the operations section; see above.

e Mairie de Paris Annual Report (2012: 39).

Appendix 4.5 Changes in population and volumes consumed

	1960	*1990*	*2000*	*2010*
Population (millions)[a]	2.79	2.20	2.11	2.27
Annual volume consumed (Mm^3)[b]	210	253	212	186
Daily volume consumed (m^3)	575,000	693,000	575,000	510,000
Consumption l/d/h[c]		231l/d/h	192l/d/h	170l/d/h

Notes

a The population figure for 1960 comes from Garnier *et al.* (1966: 29). For 1990 and 2000, Euzen (2004); for 2010, Mairie de Paris Annual Report (Public services 2012: 9).

b Source for 1960 and 1990 Euzen (2004); for 2000 Mairie de Paris rapport annuel services publics (2009: 22, "authorized volumes consumed"); for 2010 Mairie de Paris rapport annuel services publics (2012: 39). The "volume consumed" includes the volumes of service water, unbilled volumes and volumes billed to subscribers.

c The population corresponds to the legal population recorded in the census; for the calculation of consumption per inhabitant, the daytime population needs to be added: non-resident workers and tourists; Eau de Paris (Annual Report 2010: 3–4) estimates the number of daytime users at three million, including workers (who serve the tourists).This is a cautious estimate, if one believes Euzen, who estimates a figure of four million daytime users.

Notes

1 This paper was presented and discussed at the conference organized under the "Cities" programme by SciencesPo and Collegio de Mexico on 30 January (Mexico) and 26 April 2014 (Paris).

2 For the situation prior to the development of the large technical system see Dardenne (2005).

3 We also use the expression "small" institution to differentiate from "large" institutions, or instruments, or tools. For an interpretation inspired by Weber and Foucault, see Lascoumes and Le Galès (2004).

4 The changes in the terms used to describe these practices over the past 25 years reflect the difficulty of fitting this type of public action into general categories: delegation, PPP; lessor//lessee; organizing authority//operator; licensor//licensee; principal//agent.

5 This institution is the highest legal body that rules on cases where public entities are implicated; from case to case, decisions (arrêts) made by the Conseil d'État have contributed to an elaborate doctrine concerning many collective issues in France; conflicts between private entities are examined by civilian courts.

6 J.L. Trancart, "Working paper pour une water story", version 6, February 2012: 2. The company worked through a cost-plus contract, its activities evolving over time. While the start of this story, from 1860 to 1880, is becoming common knowledge the 100 years that followed are less familiar, and this industrial relationship was punctuated by a number of major events: Municipal Autonomy Act (1884), First World War, 1929 crisis, Second World War, urbanization and change in the relations between Paris and the suburbs.

7 In an extension of its long-time contract, CGE would manage billing on the left bank until 2003.

8 Talk by Anne Le Strat, "Water in Paris, back to the public!" Casuistry seminar, "The common Good, common goods, the commons, the collectivisation of interests", EHESS, Paris, 21 January 2014.

9 See law (Sapin) and decrees (Barnier, Mazeaud).

10 Le Strat, op. cit.; interview of an executive in position at that time.

11 Cour des comptes or the equivalent of the General Accounting Office (GAO) in the United States, familiarly called "la Cour".
12 Paris therefore ranks alongside Singapore, one of the international benchmarks. The data provided correspond to the report approved by the municipality and the operators, disseminated either in the 2009 price report (op. cit.), or in Issue 46 of *Aquae*. Mairie de Paris, Annual reports on the price and quality of the public potable and wastewater services, 2009 (p. 22). BIPE 2008.
13 In 2009, production stood at around 190 million m^3 a year (i.e. 520,000 m^3 per day) (see Appendix 4.5).
14 In 2011, in the documents of the Water Consulting Committee, the chairwoman of Eau de Paris explained that the price of drinking water had gone up by 260 per cent during the 25 years of the contract. As Bernard Barraqué points out (2012: 908 and n. 12), there is a mistaken use of statistics, which confuses percentage increase, index increase, current price and constant price. Arithmetically, an increase of 260 per cent means that if the price was 100 in 1985, it would be 360 in 2009.
15 Bertrand Delanoe, 5 November 2007, quoted by J.L. Trancart, p. 5.
16 Claude Martinand, Head of the "Corps des Ponts" State Engineering Corps, Chairman of the Institute of Public Service Contracting, interview *Aquae*, no. 46, p. 7.
17 Interview with a former member of staff of the first deputy mayor at the time, elected mayor in 2014.
18 Eau de Paris annual report 2011. Price of drinking water €1.0464 per cubic metre in 2010 and €0.9627 per cubic metre in 2011.
19 For average consumption of 90 cubic metres per household per year. Mairie de Paris (2012: 28).
20 *Le Monde*, 2 December 2013, counter-survey, economics, p. 14. Data produced by Jacques Friggit/CGEDD.
21 *Le Monde*, 11 and 12 December 2010, p. 16. For the beginning of the cycle, see also *Le Monde*, 21 and 22 November 2004 and the feature by Isabelle Rey-Lefebvre. In 1998, at the low point following the 1991 collapse in the property market, the average price for old property was €2,500 per m^2.
22 *Le Monde*, 13 and 14 October 2002, Isabelle Rey-Lefebvre, feature, "Property in Paris: After the boom, the crash?"
23 Paris is becoming gentrified. It is not a new phenomenon, but it is speeding up.
24 This land tax includes the €300 tax for the removal of household waste.
25 In January 2010, 250 professionals from the private operators joined Eau de Paris. *Acquae*, no. 46, November 2009.
26 In order to track the implementation of the performance contract one has to study the administrative investment budget, at the beginning and end of the contract, to identify the effective undertakings. Eau de Paris produces no annual summary.
27 Interview with a former senior municipal civil servant, January 2014.
28 Wastewater transport and treatment are handled by SIAAP, an intercommunal legal entity, whereas the fees are paid to the Seine Normandie Basin Agency and to Voies navigables de France (the body that manages France's navigable waterways).
29 Potable water emerges with a budget of €2.7 million and investments of €383,102 for the financial year 2010, well short of the full picture.
30 Earnings before interest, tax, depreciation and amortization. or gross operating profit.
31 "EBITDA is sufficient to repay the principal on the debt in 0.62 years" (Eau de Paris, 2010 report, p. 66).
32 We spoke to colleagues, operators and retired executives, and asked questions of Eau de Paris.

33 Mairie de Paris, Annual reports on the price and quality of the public potable and wastewater services, financial year 2009, p. 23. See also this argument developed by Barraqué (2012).
34 Mairie de Paris (2012: 9).
35 This price is €3.30 per cubic metre, based on a standard volume of 120 cubic metres per year, which makes comparison possible (Mairie de Paris, 2012: 27 and 28).
36 Interview with a former municipal executive, January 2014. In 2003, there was no management control over the Mayor of Paris, no procurement department. At the time, there were 30 semi-public companies, numerous associations and a few heavyweights like the Prefecture of Police, the Fire Service and RATP, poorly controlled by the finance department.
37 Talk by Anne Le Strat, "Water in Paris, back to the public!" Casuistry seminar, "The common good, common goods, the commons, the collectivization of interests", EHESS, Paris, 21 January 2014.
38 In fact the turning point is changing according the criteria taken into consideration. The year 1962 represents the maximum in production (375 Mm^3) and the year 1990 the maximum for consumption (255 Mm^3). Discussion with Olivier Jacque, former director of the water and wastewater service of Paris (2003–2010).
39 Agathe, Euzen, "Que se cache-t-il derrière les courbes de consommation d'eau? L'exemple de Paris", Communication, Latts, Ecole des Ponts, May 2004, p. 4.
40 See Chapter 5 by Thomas Blanchet (this volume).
41 Interview with a staffer of the deputy mayor concerned, January 2014.
42 Evidence of a symbolic fixation on the water sector: in the same period the municipality transferred ownership rights and public service activities in the forum des Halles, at the centre of Paris, to an insurance company and a land company. It also turned to the private sector to develop a programme for shared bicycles, and then for shared electric cars.
43 See Chapter 3 by de Gouvello and Chapter 8 by Hardy and Poupeau (this volume).

5 A partial privatization under pressure

The Berlin Water Company

Thomas Blanchet

On 29 October 1999, a few days after the start of a new legislature, the old Berlin Parliament met in extraordinary session to ratify the law on the partial privatization of the Berlin water company, Berliner Wasserbetriebe (BWB). A few hours later, the Berlin Senate finalized the sale of 49.9 per cent of BWB's assets to a consortium consisting of the German utility RWE (Rheinisch-Westfälisches Elektrizitätswerk), the insurance company Allianz and the French group Vivendi (former Compagnie générale des eaux, now Veolia). This partnership, with a planned term of 29 years, for which the city of Berlin received €1.68 billion, was to unleash local political passions and attract interest from numerous international private operators. How could the water services of Berlin, capital of a country known for its local public management and its excellent technical competencies in the water sector, have been opened up to cooperation with a private foreign operator? And why did this cooperation arouse so much opposition?

This case[1] relates on the one hand to the varieties of capitalism and to models of urban governments (Boyer, 2002; Crouch and Streeck, 1996; Czada and Lehmbruch, 1998; Hall and Soskice, 2002; Kocka, 1996; Lorrain, 2005), and on the other hand, to the instruments of public action (Lascoumes and Le Galès, 2004; Lascoumes and Simard, 2011; Lorrain, 2004, 2008a). The central thesis on models is that in countries characterized by a market economy, different configurations have emerged over long periods of time, configurations that are underpinned by specific primary principles. For example, in Germany at the end of the nineteenth century, the equation proposed was that essential public goods (*Daseinsvorsorge*) be managed by public entities. This principle was embodied in formal institutions with responsibility for the municipality and in the establishment of a specific form – the multi-sector municipal enterprise, the *Stadtwerke* – which was subsequently implemented through different instruments (or second-rank institutions) operating at the coal-face: technical and accounting standards, methods of pricing and profit sharing, monitoring rules. These principles were all shared among/used by the protagonists of the water sector. The management of urban services in Germany therefore constituted a "strong local public model" (Lorrain, 2005) espoused by the political elites, regional civil servants and a large proportion of the population. In their view, this system, which gave them direct control of the action – through public ownership – expressed a more general political principle, the principle of the free administration of the municipalities.

Against this background, the involvement of a private consortium in Berlin's water management, albeit being a minority shareholder and under public control, would seem in many respects to generate tensions between a model with its own coherence and a concept of urban public action characterized by a belief in the contracting out of services. Initially, it was the elected officials who called on the private sector to resolve problems of public management, and to cope with Berlin's increasing budget deficit. However, once the reform was underway, it was associations from civil society who expressed their objections, relying on the second-rank institutions to demonstrate that the application of certain rules (e.g. profit sharing or the procedure for calculating price increases) perverted the higher principles of public management. These "small" institutions therefore had a dual function: they provided a concrete illustration of contrasting approaches in the manner of resolving practical problems, and because they were public and familiar to all, they could provide a basis for contesting decisions: they became the instruments of controversy. This case represents not only an attempt to import an institutional solution, but also a confrontation between two models of urban services.

This chapter studies the various protagonists involved in the development of this partnership, their motives, their strategies and their interests. It begins by examining the period preceding the partial privatization, the problems that local politicians faced after Reunification and the first reform of BWB. It then shows that this partial privatization was made possible by substantial work on the institutions (tariff structure, legal form) which gave rise to a complex model of cooperation. Finally, the presentation of the partnership shows how the question of the increase in prices for water services and the non-proportional distribution of dividends among shareholders were used to remunicipalize BWB. Beyond the political debates over the benefits or disadvantages of the partial privatization, the different protagonists relied largely on the "small" institutions as a way to maintain or challenge the partnership, leading ultimately to a remunicipalization that reaffirmed the German model of water management.

The development of water services in Berlin before the partial privatization

Following Reunification, the municipal company BWB found itself facing two major problems. First, it had to invest significant sums in the network and in human resources. At the time, the priority was to extend the wastewater system to the east of the city, a process in which investment rose from €144 to €517 million between 1991 and 1994. With a workforce exceeding 7,000, BWB had also to train staff and equalize the pay conditions of employees from the East and West. In addition, BWB had to deal with a significant reduction in its activity because of the exodus of people to the West. Whereas the company's management expected Berlin's population to reach four million inhabitants in the years following Reunification (BWB, 1992: 15), it actually fell steadily to below 3.4 million at the end of the 1990s.[2] Finally, Berlin's budget deficit deepened drastically, in particular owing to the removal of the subsidies granted by the Federal Republic of Germany to West Berlin (*Berlinhilfe*) from 1992.

Against this background, Parliament began to debate reforms in the legal status of the company, which was then suffering as an autonomous municipal enterprise (*Eigenbetrieb*) from its dual political and financial dependence on Parliament. Whereas the Christian Democrats (CDU), the Social Democrats (SPD) and the Liberals (FDP) saw the conversion of BWB into a share company as an appropriate solution to the company's autonomy and financing problems, the Greens and the far left (PDS, now *Die Linke*) feared a loss of parliamentary influence over local water policy and a first step towards privatization. The plan was also perceived by the trade union Öffentliche Dienste, Transport und Verkehr (ÖTV, now Ver.di) as a threat to working conditions, to the principle of codetermination and to stable water prices, because of the tax on wastewater treatment.[3] As a compromise with ÖTV, Parliament finally decided to convert BWB into a public law establishment (*Anstalt des öffentlichen Rechts* – AöR),[4] with independent legal status, placed under the authority of the *Land* of Berlin. The latter would be responsible for defining the company's role and appointing the board of directors, which in its turn would appoint the management, which thereby obtained greater freedom in managing the company.

On 1 January 1994, a new board was appointed, with the task of managing necessary investment in the network, rationalizing the company and diversifying its activities. The board then succeeded in reducing BWB's workforce, as well as its infrastructure investment. It also regularly increased tariffs as a way of offsetting the fall in consumption (from production of 287 million m^3 in 1992 to 226 million m^3 in 1998) and in wastewater treatment volumes (from 268 million m^3 to 239 million m^3 over the same period). In 1995, for the first time since its unification, BWB made a net profit of €26.3 million. In parallel, a strategy of internationalization and diversification was launched in order to generate new revenues. Predicting the opening up of the water market to competition, the company planned to export its skills (BWB, 1996: 20). Alongside technical collaborations, BWB took part in numerous international tender processes and acquired multiple subsidiaries. BWB's diversification strategy was, however, poorly judged, and contributed to increasing the company's deficit. In 1998, whereas the business of municipal water management generated a net profit of €33 million, the subsidiary SVZ Schwarze Pumpe, acquired in 1995, on its own made a loss of €50 million.

The construction of a complex legal model

The debates on the partial privatization of BWB began in 1996 in a dual context of local elections and a public deficit perceived as increasingly catastrophic. Different solutions for privatizing BWB were therefore discussed by the parties in the governing coalition. Within the SPD, the Senator for Finance, who wanted to eradicate Berlin's deficit as quickly as possible, advocated the idea of a 25- to 30-year concession contract with a private operator, for a price potentially as high as €1 billion. Backed by the leadership of the SPD, the plan nevertheless failed to win over the party grassroots. Motivated by the long-term economic potential of the company, the Senator for Economic Affairs, Elmar Pieroth, supported by the majority of the CDU, wanted to convert BWB into a share company and sell up to 49 per cent of its assets to investors from outside the water sector. Originally developed by Roland Berger's advisory bureau, this project's aim was to attract large companies into BWB's share capital, while maintaining local control over the water services (see Box 5.1).

Box 5.1 The Senator for Finance and budget recovery policy

Following the elections of October 1995 and with the support of the president of the SPD group in the Berlin Parliament, Annette Fugmann-Heesing was appointed Senator for Berlin's finances in January 1996. Possessed of great experience in public administration and a PhD in law from the University of Bielefeld, reputed for her capacity to carry through austere budget policies, her task was to control the Berlin budget, something that none of her predecessors had succeeded in doing since Reunification. After joining the SPD in 1974, Fugmann-Heesing began her career as a municipal councillor (*Regierungsrätin*) for the town of Detmold, within the chancellery of North-Rhineland-Westphalia. In 1985, she became Treasurer for the town of Herford and Director of municipal *Stadtwerke*. In 1991, she was appointed Minister of Finance for the Hesse *Land*, where she managed to cut expenditure by 4 per cent.

Supported by the leadership of her party, she carried out her roles with a clear sense of direction: to improve the health of Berlin's budget by renegotiating collective agreements with the trade union ÖTV, to cut jobs in the public sector and the administration, to increase business taxes (*Gewerbesteuer*), and primarily to oversee the sale of assets belonging to municipal companies. She was thus involved in introducing the bill on the budget structure (*Haushalstsstrukturgesetz*) with the aim of saving several hundreds of millions of euro a year through reductions in public expenditure, characterized in particular by an increase in the dividends paid by the municipal companies to the *Land* of Berlin. She also contributed to the creation of the *Vermögensaktivierung AG* (asset activation) working group within the SPD parliamentary group, set up to organize the privatization of Berlin's municipal companies. Having orchestrated the sale of the assets of the Berlin Electricity Company (BEWAG) in 1997 and of the Berlin Gas Company (GASAG) in 1998, Fugmann-Heesing organized the partial privatization of BWB. For her, as for many senior figures in Berlin, covering the structural deficit of the Berlin *Land* by selling the assets of the municipal companies represented a significant electoral issue for the local elections of October 1999.[5] Continuing as a SPD MP in the Berlin Parliament until 2011, she was involved in other activities in parallel, in particular as an adviser on public–private partnerships at BBD Berliner Beratungsdienste, a company specializing in advice to municipalities and municipal companies.

With the support of its many political backers, ÖTV opposed the partial privatization of BWB. By organizing numerous demonstrations with the employees of Berlin's municipal companies, by putting forward alternative solutions to the partial privatization,[6] and by threatening to refer the matter to the Constitutional Court under Article 24 of the Constitution, which recommended a ban on the privatization of local monopolies, it succeeded in imposing a number of conditions (maintenance of BWB's public status, a 15-year ban on employee redundancies) and in substantially altering the privatization policy (see Box 5.2).

Box 5.2 An atypical privatization model

Inspired by the model of privatization employed for the Berlin Regional Bank (Landesbank Berlin) in 1993, the holding company model was developed by the international law firm Freshfields Bruckhaus Deringer. In an article published in a European law journal, the lawyer Benedikt Wolfers, a partner in the firm, presented the legal foundations of this model, which was to set a precedent in the water sector (Wolfers, 2004). While Berlin municipality had formally sold 49.9 per cent of the assets of the holding company – of which BWB was a wholly owned subsidiary – to private investors, this shareholder structure was only made possible by a complex legal construction, based in the fact that the municipality wanted to partially privatize BWB while keeping the company under a public law regime. In order to reconcile these two requirements, at first sight legally incompatible, the lawyers employed a specific form of shareholder structure – atypical tacit shareholding (*Atypische Stille Beteiligung*) – which allows a silent partner to invest capital in a company without participating in its management. The partner receives an appropriate return on the capital invested in the company and also shares in its losses. The advantage of this form of shareholding is that it applies regardless of the company's legal status.

In the case of BWB, the municipality created a private law holding company which became a silent partner, holding 49.9 per cent of BWB's capital through an atypical tacit shareholding agreement. Through a second atypical tacit holding, the investors financed 100 per cent of the 49.9 per cent of the first shareholding between BWB and the holding company. Alongside this arrangement, two contracts defined the distribution of tasks among the partners and therefore enabled the private operators to be directly involved in the management of the company. A contract between BWB and the holding company allowed the latter to give instructions to BWB's management. A second contract between the holding company and the municipality gave the holding company the right to appoint the members of the supervisory board. These contracts provided for a Senator to appoint the chair of the supervisory board and for one of the private operators to appoint the chair of the board of directors. Finally, the investors obtained 49.9 per cent of the holding company's assets (Wolfers, 2004: 117).

On 7 July 1998, the CDU and the SPD reached an agreement to partially privatize BWB by means of the holding company model, whereby part of the capital could be sold while maintaining BWB's public status. A steering committee, consisting of three Senators (economics, finance and environment), and alongside the leaders of the CDU and SPD parliamentary groups, was then created to orchestrate the partial privatization. To this end, the committee surrounded itself with a plethora of consultants to help formulate the law on the partial privatization of BWB (Beveridge, 2012: 57),[7] in particular with regard to the calculation of service prices and the rate of return paid to shareholders, the negotiations with the private investors and the drafting of the contract. In addition, the committee had to negotiate the establishment of a formal agreement with ÖTV and obtain parliamentary approval for the bill on the partial privatization of BWB. After being rejected at first reading on the grounds that it reflected almost entirely the interests of the Senate's Finance Department, the bill was adopted on its second reading on 29 April 1999.

In order to stop the privatization, in May 1999 the parliamentary opposition referred the bill to the Berlin Constitutional Court, which approved it on 21 October 1999, subject to amendments, in particular relating to the returns paid to shareholders. The Court ruled that the rate of return on capital, at "R + 2%",[8] was too high relative to the risks taken by the investors, and rejected the "+2%". To pre-empt such a situation, the steering committee had already introduced Clause 23.7 into the privatization agreement, in order to provide the private partners – through a non-proportional distribution of the dividends between shareholders, and as a final resort (i.e. company deficit) a levy on Berlin's budget – a return on the basis of "R + 2%" in the event of the tariff clauses being amended. Right up to the signature of the contract, Vivendi apparently argued for BWB's selling price to be renegotiated downward (Schulte, 1999b: 32). However, under pressure from the Senator for Finance, who was keen to cover the 1998 budget deficit, the selling price was maintained and the contract finalized only a few days after the bill was approved by the Constitutional Council.

At a time when there was a trend towards public service deregulation and the privatization of water services, Berlin represented one of Europe's biggest contracts and a strategic bridgehead for companies wanting to develop their activities in Eastern Europe.[9] In January 1999, after shortlisting, six applicants were chosen from the 17 submissions: (1) Eurawasser (a subsidiary of Lyonnaise des Eaux and Thyssen-Krupp, and Bankgesellschaft Berlin; (2) RWE and Vivendi; (3) Enron and Landesbank Hessen; (4) Severn Trent; (5) Saur; and (6) Bewag and Preussenelektra. As this list shows, the contract attracted investors who were looking to diversify, such as Preussenelektra and Gelsenwasser, the subsidiary it acquired in September 1998, or the gas company Enron with Azurix (Defeuilley, 2001). Since many local players were reluctant to cooperate with a company specializing in the water sector, since they wanted to retain operational control over the company, the Berlin contract thus represented a development opportunity for those investors.

For the foreign investors, cooperation with a German company seemed a necessary step to strengthen their bids and gain acceptance from local players. Conversely, a lack of partners or the withdrawal of a partner during the competition constituted a serious handicap. In March 1999, Thyssen-Krupp quit the consortium, obliging Lyonnaise des Eaux to look for a new partner and to reformulate its bid. Severn Trent's difficulty in finding a partner would also seem to have compromised its chances of winning the contract (Schulte, 1999a: 35). By strengthening their political relations with the different local players, the investors hoped not only to increase their chances of winning the contract, but also to reinforce the acceptance of privatization among some of its opponents. Some of the companies were able to exploit existing professional relationships, such as Vivendi or Severn Trent, which were already cooperating with BWB internationally. Others, like RWE or Lyonnaise des Eaux, employed lobbyists to try to gain the sympathy of political groupings, BWB's management or the trade union ÖTV.

Apart from financial interests, compliance with the procedure would seem to have been an important factor in the decision-making process, as shown by the procedural irregularities perpetrated by two of the finalists. In December 1998, the press announced that Eurawasser, the favourite of the Senate's Finance Department, had obtained the memorandum drawn up by Merrill Lynch before all the other bidders.[10] This affair, whether true or not, would seem to have ruled out any chance of the candidate winning the contract.[11] In March 1999, Enron was also discredited by a conflict of interest involving Merrill Lynch, which had won the contract to manage the stock market flotation of Azurix (its new subsidiary for the water business). This cast doubt on the investment bank's neutrality in the choice of BWB's future partner, insofar as Enron-Azurix's winning of the contract would alter its initial stock market value and therefore Merrill Lynch's fees for the transaction. The parliamentary opposition and the union ÖTV called for the tender process to be cancelled. The two other applicants threatened to make a complaint if Azurix were to win the contract.

Finally, the complexity of the holding company model was a significant obstacle for many applicants. Some, like SAUR or Thyssen-Krupp, quickly pulled out of the contest after becoming aware of the conditions of cooperation. Others began negotiations with the steering committee to propose an alternative to the privatization model. For example, Preussenelektra proposed to acquire only the competitive part of BWB for €1.18 billion, thinking that privatization of the water services would inevitably lead to a rise in water prices. Even Enron, one of the finalists, doubted the viability of a privatization model in which BWB was to remain 100 per cent public.[12] By contrast, the consortium headed by Vivendi made no serious attempt to question the cooperation model, which had already proved its worth in the privatization of the Bank of Berlin.

On 18 June 1999, the steering committee awarded the contract to the consortium consisting of Vivendi, RWE and Allianz,[13] which had employed numerous consultancy firms, in particular KPMG, Crédit Suisse, Freshfield and HPC (an environmental consultancy firm), as well as around 80 consultants, at a total cost of €15 million (Beveridge, 2012: 61). The consortium undertook to comply with numerous conditions imposed by the steering committee, notably:[14] to hold water prices stable until the end of 2003, and to reduce them in the long term; not to conduct any redundancy plans for 15 years and to create jobs; to guarantee BWB's economic independence; to invest at least €2.56 billion in infrastructures over ten years; to develop BWB internationally; and to foster technological research and development in the water sector in Berlin. After signature of the contract, the company paid Berlin €1.68 billion, including €102 million for the restructuring of the firm SVZ Schwarze Pumpe. A few days later, a new board of directors was appointed, with a Vivendi manager as Director of Operations, a RWE manager as Director of Finance, and two BWB managers as Technical Director and Director of Human Resources.

The partial privatization: a tricky coexistence?

Against a continuing background of falling water consumption (from 216 to 192 million m^3 for consumption and 219 to 204 million m^3 for wastewater between 1999 and 2009), practical action by the operators, apart from massive investment in water research in Berlin[15] and in efforts to improve customer relations,[16] was characterized by significant work on the technical system and on operational reorganization. First of all, a major organizational change took place. Local water management was separated from the other activities, which were transferred to the holding company and divided into four departments: operation (local water management), international, services and multi-utility (BWB, 2000: 8). The activities associated with diversification and internationalization were mothballed or suspended, and a number of subsidiaries sold. By 2006, the holding company had only three departments: international, regional (local water management) and services. With a 15-year moratorium on redundancy plans, the management introduced a large-scale incentive programme for early retirement and part-time working. Employees were transferred to the holding company's numerous subsidiaries. Others were trained to perform alternative roles within the company. By avoiding the need to recruit new staff and to employ external service providers, this policy sought partly to offset the loss of revenue caused by the price freeze over the first four years.

Under the privatization contract, the operators had undertaken to invest €255 million per year in the refurbishment and extension of the water networks. In practice, investment in the system between 1999 and 2009 turned out to be higher than this. Up until 2002, BWB essentially invested in extending the wastewater system, while the subsequent years were marked by investment in network renewal. From 2004, an automation programme for the 143 pumping stations was introduced, which also helped reduce the workforce (BWB, 2004: 35). The consortium would therefore seem to have fulfilled many of the undertakings set out in the privatization contract. Despite this, it was unsuccessful in reducing long-term water prices.

The purpose of the political decision to freeze water service prices for four years was to make the partial privatization more acceptable to the local population. At the end of the period of price freeze (2003), the Senate introduced two amendments to the bill and to the contract in compensation for the clauses that were declared unconstitutional in 1999. These changes were ratified on 11 December 2003 by Parliament, which had been dominated since 2002 by a coalition between the SPD and Die Linke. First, the interest rate R + 2 was replaced by a regulatory interest rate set each year by Senate decree, which rose from 6 per cent in 2004 to 7.77 per cent in 2008 (Hüesker, 2011: 258). Second, the method of calculating depreciation was no longer based on the acquisition value (purchase price at the time of acquisition), but on the replacement value (amount needed to obtain the same product if it had to be replaced on a defined future date), which took account of changes in the inflation rate (see Box 5.3).

Box 5.3 BWB's pricing structure

At the end of 2007, under pressure from consumer associations, BWB published a brochure giving details of the pricing structure for water services.[17] This pricing structure was calculated annually on the basis of the total costs generated by water distribution and wastewater treatment, divided by the quantity of water distributed or treated. On the one hand, the fixed costs included spending on equipment and staff, spending on the raw materials and consumables used, and miscellaneous costs. Those costs, which were the biggest component of the pricing structure, were not a matter of controversy. In contrast, the assessed costs, including depreciation (the phasing of the cost of investment over the term of its use), assessed interest and the miscellaneous costs, were at the core of the conflicts, because they were the basis for the rate of return paid to shareholders.

The change in these two rules and their consequences gradually came to be exploited by the different protagonists to feed or mitigate the controversy around the partial privatization. While consumer associations and MPs condemned these repeated increases as being caused by the excessive returns paid to the shareholders, the latter tried however to qualify the impact of the partial privatization upon the increase in the prices of services. For BWB's management, the increase was associated with a general fall in water consumption. For the private operators, after seven years of frozen prices, it was inevitable that there would be increases[18] (which were, moreover, much smaller than before the partial privatization) (Oelmann *et al.*, 2011: 3; Schwalbach *et al.*, 2011: 6). In 2003, the Senator for Economic Affairs announced that the regulatory interest rate set for 2004 would be much lower than the national interest rate ($R=6.2$ per cent).[19]

The increase in prices and the unequal distribution of dividends[20] triggered significant tensions within the supervisory board. One notable weakness of Berlin's local government at the time lay in the divergence of interests between the Senator for Finance, who favoured the price increase for budgetary reasons, and the Senator for Economic Affairs, who opposed it. As a result, the private partners enjoyed the former's backing in the annual negotiations on the dividends to be paid to the shareholders. Nonetheless, in 2006 the private operators claimed €340 million in compensation from Berlin under Clause 23.7 of the partial privatization contract. In 2008, the shareholders instituted an arbitration procedure to settle this dispute. The Senator for Finance then joined forces with the Senator for Economic Affairs. At the same time, a divergence of interests began to emerge between the two private partners. Whereas Veolia had no wish to leave Berlin, RWE wanted cash in order to refocus on the energy sector and decided to sell its shares. It was therefore opposed to a dividend reduction, which would tend to lower the share price.[21] Under pressure from RWE, the arbitration procedure continued, and in 2012 the lawyers ruled in favour of the private investors (see Box 5.4).

The case was then taken up by the Cartels Office, which – on the basis of the ratio between costs and revenues – compared the prices of water distribution in Berlin with Munich, Hamburg and Cologne, and in December 2011 imposed a 19 per cent cut in water prices for the next three years (i.e. a €250 million reduction in BWB's revenues out of annual revenues of around €1,170 million). The company's management then instituted a legal procedure to show that BWB's tariffs were legally regulated by the government and were therefore closer to taxes, a domain in which the Cartels Office had no authority to intervene. It also criticized the Office's comparison for failing to take into account the specific situation of Berlin post-Reunification. In April 2012, the Office reiterated its demand, requiring a 21 per cent cut in water prices below the 2010 level, for the next four years (i.e. a €292 million reduction in BWB's revenues) and called for the company to take action. In response to this demand, the company's management referred the case to the Düsseldorf Court of Appeal.

Box 5.4 The Senator for Economic Affairs

On his arrival in Berlin in 1977 to complete his studies in political science, Harald Wolf was already active in different political groups with Trotskyist affiliations, and became successively involved in different political parties (Alternative List for Democracy and Environmental Protection, Socialist Left – *Sozialistische Linke* – the Greens) before joining the Berlin PDS (Social Democratic Party). In 1991, he entered the Berlin Parliament and became chairman of the PDS group (now Die Linke) from 1995 to 2002. At the time, he was one of the most fervent opponents of the partial privatization of BWB and one of the joint instigators of its constitutional investigation.

In 2002, Wolf was appointed Senator for Economic Affairs, Technologies and Women's Affairs for the *Land* of Berlin, in which role he became chair of BWB's supervisory board. Responsible for setting the interest rate used as a basis for shareholder returns, he was criticized for not putting a brake on water prices. Despite this, he introduced a number of measures designed to limit the fall in consumption. In 2007, he succeeded in modifying the pricing structures for water (shifting from fixed pricing to a combination of a base price and a consumption-based variable price) to reduce water costs for large consumers. In parallel, in 2008 he introduced an obligation of use (*Benutzungszwang*) prohibiting companies from digging their own wells. During his mandate, he set up an office that monitored and planned the development of Berlin's public companies. In the case of BWB, this office also aimed to support the Senator for Economic Affairs in its work with the private partners. Restricted by the rules for the partial privatization, the Senator had only limited influence on water pricing. In March 2010, therefore, he referred the prices charged by BWB for investigation by the Cartels Office. By this means, Wolf reckoned to achieve a long-term cut in water prices of 15 to 20 per cent by getting around Clause 23 of the privatization contract. Despite his pro-remunicipalization stance and his attempt to influence the prices of water services, he declared himself against the citizen initiative for the publication of the contract and the remunicipalization of BWB, on the grounds that the Senate had already published the contracts.

From civil opposition to the legal control of contracts

The controversy over prices and the distribution of dividends also extended outside the sphere of the shareholders, as a coalition against the partial privatization was formed. Supported by associations of anti-globalization activists (Attac), consumers, tenants or taxpayers, as well as by MPs from different factions (Die Grünen, Die Linke, SPD), a citizen group was formed in May 2006 to organize a referendum to force publication of the privatization contract. Officially constituted in 2007 and given six months to collect 20,000 signatures from inhabitants registered in Berlin, the initiative had obtained double that number by the required date. The Senate tried to block the measure by rejecting the citizen request, but the latter was approved by Berlin's Administrative Court on 6 October 2009. The initiative then had to garner a further 173,000 signatures by October 2010, by which time it had collected 280,000, enabling it to organize the referendum on 13 February 2011. In order to make the initiative superfluous, the Senate amended the Freedom of Information Act to allow the contract to be published. Despite this, the referendum went forward with the aim of obtaining ratification for a citizen law to deny legal standing to unpublished documents and to prevent the partnership from obtaining legal protection through the use of secret documents. For this to happen, 615,000 registered inhabitants (a quarter of the electorate) had to vote "yes" in the referendum. On 13 February, more than 666,000 people voted for the contract to be published. In March 2012, Veolia and RWE filed suit with the Federal Constitutional Court to prevent new documents (emails, letters and internal decisions) linked with the contract from being made public.

The publication of the contracts led to legal examination, at the heart of which was Clause 23.7. It enabled opponents to employ different formal rules to try to invalidate the partial privatization contract. In March 2011, at the instigation of the environmental group Grüne Liga, a panel of lawyers sought to show that the guarantees granted to the private partners ran counter to Berlin's budget law. In June 2011, the Berlin consumer association and Transparency International filed a case in Brussels to prove that the payments made to the private partners were akin to the state subsidies prohibited by the European Commission. Under the impetus of the citizen vote, a parliamentary committee, consisting of nine experts (three SPD, two CDU, two Greens, one Linke, one Pirate), was formed in December 2011 to discuss ways of legally invalidating the partial privatization contract. Although the different attempts to employ legal avenues did not directly contribute to the invalidation of the contract, they had significant influence on the remunicipalization of BWB.

During the local parliamentary elections of September 2011, the remunicipalization of BWB became an even more central political issue. While the SPD and Die Linke were calling for remunicipalization, the CDU and FDP parties favoured a tariff reduction negotiated with the private partners. However, BWB's management refused to consider this as long as the dispute with the Cartels Office had not been settled. The citizen initiative was also opposed to remunicipalization at any cost, because of its potential to damage Berlin financially, and preferred to wait for the result of the dispute with the Cartels Office.

At this time, the Senator for Finance, Ulrich Nussbaum, was already negotiating a buyback of RWE's assets. In April 2011, the energy company put forward an initial proposal of €800 million, too high for Berlin. In May 2012, RWE and Berlin agreed on an amount of €650 million, which was approved by Parliament in October 2012 at the same time as a €60 million cut in water prices for the year 2013. Soon after, Veolia announced that it would not preclude selling its assets on the same terms as RWE. Alongside the price of water, control of the company represented a real issue for both Berlin and Veolia. Although it owned 75.1 per cent of BWB, the local government, because of the framework contract established in 1999, could neither take charge of operations nor appoint BWB's chief executive. Veolia announced that it had no interest in continuing cooperation if it was not able to retain operational management of the company. At the end of 2013, the full remunicipalization of BWB took place. The Senate announced that it had reached a compromise with Veolia to buy back the assets for €590 million, an agreement approved by

Parliament in November, though without the votes of the opposition, which felt that such a project would not achieve a long-term reduction in water prices. On 2 December 2013, Veolia and the Senate's Finance Department finalized the buyback of BWB, making it once again an entirely public company.

Conclusion

The Berlin case shows how the "strong local public model" of urban services in Germany held out against the establishment of a partnership with private operators. This is evidenced by the intense political debates on the introduction of the private sector and focusing on pricing, legal models, evaluation of the company's assets. It is further evidenced by the tricky coexistence among the different protagonists and the recourse to a number of legal rules to increase pressure on the partnership, which finally put paid to cooperation. In this case the protagonists have mobilized – from the beginning, in 1999, to the final act in 2013 – many "small" institutions to support their view. In fact, the company's work on the water services does not seem to have been the core of the problem. The private operator adapted to difficult conditions, characterized by a 12.5 per cent fall in consumption between 1999 and 2009, and high expectations among the local decision-makers. Over the same period, the 26.5 per cent increase in water prices made a bad impression: the tariff hikes were only introduced after 2004, for regulatory reasons, and at a time when the unequal share of the dividends between the shareholders was starting to become a salient issue. All these factors generated heated opposition to the partnership and exacerbated the divergences of interests among the shareholders. The keenness of the local decision-makers to finalize the contract as quickly as possible by compensating the private operators through the introduction of a technical fix – Clause 23.7 – reveals the importance of the initial conditions in the success of such a partnership. This error in financial engineering led to an infringement of the principle of equality of dividends between shareholders, an essential symbol of a lasting partnership.

Notes

1 This study is drawn from a doctoral thesis prepared at the Freie Universität Berlin and at Berlin's Marc Bloch Centre. It is based on some 30 interviews conducted between 2008 and 2011 and a study of documents mostly consisting of articles in local and specialist newspapers and of minutes of local parliamentary debates.

2 *Zeitschrift für amtliche Statistik*, 3, 2010: 26.

3 In order to encourage the municipalities to keep wastewater management under direct public control, usually in the form of a municipal department, German law guarantees a tax exemption for this service, provided that it is managed by a public status organization.

4 A legal status close to the public industrial and commercial establishment (EPIC) in French law.

5 Interview, former SPD MP in Berlin, 22 February 2011.

6 In particular a concession agreement between Berlin and BWB, or else the grouping of municipal companies (water, waste, dock services, public transport) in order to generate substantial economies of scale and functional synergies.
7 Merrill Lynch was responsible for the financial and competition aspects, Hengeler Mueller, a German law firm, for the legal aspects, and the accountancy firm BDO Deutsche Warentreuhand AG for the valuation of BWB and for the pricing of water services.
8 The interest rate R generally varied between 6 and 8 per cent.
9 Interview, Vivendi consultant in the tender process, 22 April 2010.
10 Schulte (1999a: 35).
11 Interview, BWB press representative, 29 April 2009.
12 Interview, former CDU Secretary of State for Economic Affairs, 6 May 2010.
13 In 2002, Allianz sold its shares in equal proportions to its two partners.
14 Berlin Senate, press release, "Teilprivatisierung Wasserbetriebe: Senat gibt Zuschlag an RWE/Vivendi", 18 June 1999.
15 In 2001, the consortium set up the Berlin Water Expertise Centre and undertook to invest in it some €5 million a year for ten years. It also funded a chair of urban water management at Technische Universität Berlin.
16 Establishment of a call centre, simplification of the procedures for contacting the company, dedicated support service for large consumers, etc.
17 BWB, *Grundlagen der Tarifkalkulation*, 2007, p. 4.
18 www.veoliawasser.de/content/fragen-und-antworten (website consulted on 4 November 2013).
19 Berlin Parliament, thirtieth sitting of the legal committee, fifteenth legislature, p. 10.
20 See above: Clause 23.7 of the privatization agreement had established a non-proportional distribution of dividends among shareholders to the benefit of the private part.
21 Interview, Director of the Berlin Office for the Oversight of Public Law Establishment, 7 May 2010.

6 Corporate social responsibility and public service management in Nagpur

Marie-Hélène Zérah and Cécile Renouard

During the 1990s, when most of India's towns and cities still lacked sufficient access to water and sanitation, efforts were made by the World Bank and certain policy-makers to impose the large concession model upon the country (Zérah, 2003). There were a few failed attempts to introduce concession contracts and, as elsewhere, these successive failures were attributed to the unsustainable economic viability of these contracts in risk countries, as well as to the poor credibility of these contracts and regulatory bodies (Harris *et al.*, 2003), thus fuelling opposition to privatization. The use of public–private partnership arrangements for improving water supply appeared to have been abandoned, despite international experience showing it was possible to fine-tune contracts and develop policies in incremental steps that focused in particular on the transfer of expertise (Marin and Izaguirre, 2006). Nevertheless, since the beginning of the years 2000, an ambitious urban policy scheme, the Jawaharlal Nehru National Urban Renewal Mission (JNNURM), has been channelling major investment into fulfilling various urban sector reform requirements and fostering the use of public–private partnerships (PPPs) for urban services. Although they remain controversial, a few of the experiences deemed successful by the public authorities have shown that it is possible to improve the water service and provide a continuous 24-hour service.[1]

Of the most recent contracts, much attention has been paid to the contract for the city of Nagpur, as it covers a long period of time – 25 years – and aims to provide a universal 24-hour service. In addition, the service delegation contract was awarded to one of the sector's largest international consortiums, Veolia. While this project has so far been a success, the public authorities considered that this success would be consolidated by improving the range of services on offer. Long-term success will also depend on the project's ability to help develop local urban capitalism via a learning process based on the sharing of international expertise and by proving to Indian companies that water supply is a financially viable sector. For Veolia, and the private sector as a whole, success is contingent upon the opening up of the market. As projects in Cochabamba (Bolivia) and Tripoli (Lebanon), or the initial phase of PPP in India, have shown, failures can result in the market being closed for a long time (de Gouvello and Fournier, 2002; Verdeil *et al.*, 2009). Lastly, over and above its contractual obligations and as "a supportive and committed group", a number of Veolia's activities form part of its social responsibility agenda (Veolia Sustainable Development Report, 2012). As such, it is vital to establish whether, when operating alongside the public service, "delegated urban services" can help ensure that services are developed fairly.

In order to do this, it is first necessary to understand the effort that went into defining and implementing these modes of service delivery in urban areas (Jaglin and Zérah, 2010). The premise presented here is that the specific power relations found within the contracting company also influence contract implementation. A number of authors have already studied this aspect, particularly in relation to the water sector in China (Lorrain, 2010) and the Philippines (Wu and Malaluan, 2008). However, for India, analyses of the few service delegation contracts implemented have only ever considered the contractor as a single homogeneous body and this from the purely institutional perspective of the delegating authority. In contrast, the aim here is not to review the contract preparation but to gain an understanding of the company's operations by focusing on its subcontractors and licensed plumbers, and thus determine how deeply the private company is entrenched in its relationship with the state. Through establishing the porous nature of the public and private spheres, it becomes possible to assess how well a public service can be provided by a multinational displaying a commitment to corporate social responsibility.

The first part of this chapter looks at the questions posed in relation to the water sector by the literature on joint ventures and social responsibility. Next, implementation of the partnership in Nagpur is analysed by reviewing subcontractor and plumber involvement in installing connections to the water network and reveals the difficulties inherent in reconciling localized issues and overall responsibility. Information for this analysis was obtained through a field survey on the water supply methods used in poor districts of the city and on the opinions of activists and local associations through interviews (Bommier *et al.*, 2012). Additional insight was also obtained through a series of qualitative interviews that were conducted with employees of Veolia and of other companies involved in partnerships in other cities.

Partnerships and social responsibility

Operating standards for joint ventures and corporate social responsibility (CSR) were initially developed for the manufacturing sector and industries involved in exploiting natural resources. Thus, are they relevant for a sector in which the concept of public service plays a central role and where industrial activity is embedded in the relationship with the state and social organizations? The work on joint ventures focuses on the success factors involved in managing private–private partnerships. Three main factors were highlighted.

First, the way in which the partnership's ownership and control structure is set up is extremely important. In theory, compared to a joint venture where there is equal ownership, a partnership in which one partner owns a greater share of the capital could adversely affect confidence and generate more conflict (Luo, 2009). Intrinsically linked to the ownership setup is the concept of control. However, control within the partnership is not entirely dependent on the ownership structure alone, but also on the investment made in other types of resources, particularly staffing (assigning senior management to the partnership) and specific areas of expertise (local knowledge, technical skills) (Child and Yan, 1999). To a large extent, the success of international joint ventures is determined by the way in which both strategic control and operational control are managed.

Second, the two partners' motives and objectives need to be properly aligned to ensure activities are managed harmoniously and effectively. When analysing a large quantitative survey carried out in China, Steven Si and Garry Bruton (2003) found that there were extremely high levels of frustration among multinational companies. Through their work to identify partners' motives, they reveal that multinationals enter into joint ventures with two main objectives: to learn more about local organizational practices and to improve their knowledge of local issues. In contrast, Chinese companies are predominantly interested in developing a strategic market advantage or a point of entry into other sectors. Consequently, there is an initial disconnect that carries over into the partnership's operational management.

Third, a partnership developed over the long term between two parties who know each other well is also a success factor, and one which may help better align objectives (Luo, 2009). This close relationship is particularly important in joint ventures between companies from developed countries working in countries where "cultural distance" plays a key role. This cultural distance also encompasses the various interpretations of corporate responsibility and how these differ, with international organizations always attaching greater importance to the concept of corporate social responsibility (CSR) (d'Iribarne, 2009) (see Box 6.1).

Box 6.1 CSR between soft law and hard law: contrasting approaches

In 2001, the European Commission's Green Paper, *Promoting a European Framework for Corporate Social Responsibility*, defined CSR as a series of voluntary commitments that go beyond regulatory and legal requirements and are likely to give companies a strategic advantage. In the European Union's updated document of October 2011, CSR is no longer defined as a series of rules that supplement regulatory requirements but rather as the company's management of the various impacts of its activity, which considerably changes its scope. The concept of voluntary action is replaced by that of impact management, with CSR being defined as "the responsibility of enterprises for their impacts on society".[2] In the same vein, the May 2011 OECD *Guidelines for Multinational Enterprises* highlights the need for companies to take the views of other stakeholders into account and that they should "identify, prevent and mitigate actual and potential adverse impacts" linked to their operations and exercise responsibility to their supply chain, particularly subcontractors and suppliers (OECD, 2011: 22).

The approach set out by the European Union and OECD for implementation by member countries and their international companies contains far more precise guidelines than those found in recently enacted Indian law (the 2003 Companies Act). This stipulates that 2 per cent of companies' net profit is to be allocated to social responsibility activities; however, there is a risk that this will serve only to encourage charitable activities rather than foster changes in practice.[3]

Given the current international agreement on sustainable development goals and the right to water,[4] multinational companies are now being asked to explain what they are doing to help implement these rights and this from two perspectives. On the one hand, companies are expected to provide goods or services that are useful to society while sustainably protecting natural and human ecosystems, and, on the other, companies working in the water sector also have a particular responsibility to the public authorities, from whom it has a specific remit. From an overall and theoretical perspective, there is a certain ambiguity in the way corporate social responsibility is defined in international discourse. Although values such as fairness and inter- and intragenerational solidarity are upheld and companies are strongly encouraged to take on a social and environmental role, at the same time this element is presented as being compatible with ensuring financial viability or maximizing short-term profit. This is a well-known business ethics argument, whereby "*good ethics is good business*" – and whereby ethics can generate profit (Solomon and Hanson, 1985). However, in the water sector, it is difficult to align these two goals due to the lack of guaranteed financial viability of the management contracts, some of which have ended in failure. In addition, companies, aware of the risks to their reputation are coming under increasing scrutiny from both international and local anti-privatization movements.[5]

With regard to global public goods such as water, the responsibilities assigned to companies are particularly important. From an economic and financial standpoint, they need to provide universal access to the resource under their management; whereas, from a political perspective, it is necessary to ensure that sustainable management conditions and resource sharing are put in place. As local and national public authorities are also responsible for ensuring that the population has access to the water resource, these responsibilities are increasingly being taken on by companies under "shared governance" models in which a number of different stakeholders are involved. The concept of shared governance is today used to define the approach under which various "stakeholders" (including the local population) are involved in defining common goods service strategy and project management, thus drawing upon Elinor Ostrom's work on natural resources (Ostrom, 1990). There are undoubtedly specific issues at stake with regard to social responsibility in the water sector which "suggest that private companies in the water industry need concepts and processes that produce integrated approaches to address the many issues with the different stakeholders (and local communities and public authorities in particular)" (Tsogas, 2011: 231).

Literature on international joint ventures makes little mention of the importance of this "polycentric governance", which involves managing the relationship with the municipality on the one hand and the relationship with "civil society" on the other. In principle, in traditional joint ventures, the role of the international partner is specifically not to manage the market and distribution relationship and they select a local partner precisely to avoid involvement in this last segment of the production chain (Meschi and Riccio, 2008). However, this does not apply to the urban services sector, as the relationship with the customer is central to the service management company's operations. Moreover, water supply is not only a political sector; it is also a low-tech sector that makes it possible to introduce imitation processes such as the attract/extract strategies used by Chinese companies. These companies are intrinsically linked to the state, which facilitates access to the municipalities and makes them essential partners (Lorrain, 2010). Is such a model applicable to other countries, such as India, where the democratic system is more robust and highly competitive?

The company and "the state": formal cooperation and informal resistance

The Nagpur Project

Although many infrastructure projects in India have encountered delays and failure, the service delegation project in Nagpur was implemented relatively quickly following a successful pilot in one of the city's districts. Conditions conspired to create an enabling environment. First, both the central government and the government of Maharashtra allocated considerable public investment funds to the project (3,860 million rupees or around €55 million). National urban policy requires this new investment to be financed by the national, regional and local governments, with local governments having to provide 30 per cent of the investment funding. Thus, by using a private company, Nagpur municipality absolved itself of the need to raise its share of the funds. The city of Nagpur also successfully implemented a pilot project to assess the feasibility of continuous service, the contract for which had already been awarded to Veolia. Following this, the programme to expand the project to cover all of the city's 2.5 million inhabitants was championed by the Municipal Commissioner,[6] head of the executive, and politically supported by the then president of the BJP, Nagpur being a BJP stronghold (the BJP or Bharatiya Janata Party was the second largest political party in India behind the Congress Party and came to power in 2015 at the national level).[7] This agreement between politicians and bureaucrats and between regional and local levels is rare in India and, in this case, resulted in the timely launch and completion of a tender process (see Box 6.2). The tender was won by

Box 6.2 Successful project setup: the Municipal Commissioner's view

During an in-depth interview,[8] the Municipal Commissioner, who initiated the Nagpur project, highlighted the economic, institutional, professional and personal factors underpinning the success of this first city-wide delegation contract. Aware of both the investment requirements and limited municipal budget, he opted to use a multi-step process: "my initial idea was to design a project with a rapid return on investment by funding it through electricity savings".

He then set up tripartite agreements with the Department of Irrigation and the thermal power plant to sell them the wastewater, prior to taking advantage of the opportunities made available through urban policy (JNNURM). He thus underlined the importance of ensuring that there is an appropriate financial setup in place to offset budget constraints. He also stressed the importance of gaining support from local elected officials and of his communication-based method (regular meetings, media involvement, requesting engineers to organize discussions in distribution areas), as well as of reducing conflict between elected officials:

> "I endeavoured not to take any personal credit. Most of the time, I told them to announce the projects themselves and I positioned myself as leader of my engineers and as the upholder of impartial governance". To gain the elected officials' trust, he also used the innovative approach of making them accountable and forging good working relationships: "I gave discretionary payments to the elected officials to be disbursed in coordination with my area engineers ... when I was in charge of the district, I implemented funding allocated to members of the national parliament and regional assembly. So that gave me this idea, which helped as they felt they were involved in the decision-making process. I gave a presentation and everyone was surprised". His field experience of working in rural areas was an advantage: "When you work in the field, for every good idea you try to implement, sometimes it works and sometimes it doesn't. You need to try to understand their way of thinking [i.e. rural elected officials]. You need to convince them and, most of the time, they realize that it's the right solution. It is also necessary to work out how they can help you as they know the area and can come up with good ideas".

As far as his own personal career and role in the project are concerned, he highlights the importance of his knowledge of the issues and his experience: "I worked in Nagpur as a District Magistrate for three years prior to being appointed Municipal Commissioner, so I was very aware of water issues". Being a civil engineer and graduate of one of the top engineering schools has also helped: "I could understand what could be done and understand the technical reports ... yes, that gave me bargaining power ... [smiling] ... X [the director of Veolia] told me that I was a tough negotiator".

Orange City Water Limited, a joint venture between Veolia and a local company set up in 1992, VIL (Vishvaraj Infrastructure Limited), which specializes in construction (roads and bridges). The joint venture is equally owned by Veolia and VIL and the delegation contract is between Orange City Water and the municipality.

The contract involves rehabilitating the existing network and expanding service coverage to all within a five-year period and then managing the service for a further 20 years.[9] The aim is to show that it is possible to provide a good quality 24/7 continuous service. Targets for reducing unaccounted-for water were also set, requiring greatly improved operational efficiency. The operator's income is based on bills collected. The municipality retains control of the assets, as well as overall responsibility for water supply and pricing and customer connection decisions. The challenge is twofold. First, with very little investment having been made in the service over the past few decades, the consortium needs to upgrade the infrastructure, install water meters and rehabilitate the entire water distribution system. Second, like many emerging large cities, many of Nagpur's inhabitants live in slums or illegal settlements. There are estimated to be over 400 slums in Nagpur, housing nearly 910,000 people with limited capacity-to-pay (Bommier *et al.*, 2012). Thus, in managing this project, Orange City Water needs to focus on rapidly improving not only operational efficiency, but also the social acceptability of the project and the solutions implemented in the poorest districts.

The relationship with public institutions: mutual distrust over the "human resources" issue

A priori, the usual issue of integrating staff was not a problem here, as very few municipal employees wanted to join Orange City Water. The terms and conditions for transferring to Orange City Water were not advantageous as not only did staff have to undergo a trial period but, at the end of the transfer process, they also had to permanently relinquish their municipal employee status. Moreover, the municipality had enough posts vacant elsewhere to redeploy its staff to other departments. In concrete terms, this mass refusal to transfer to the consortium is a clear indicator of strong internal opposition to the service delegation project, which raises major issues for the operational delivery of the contract.[10] For example, the private company and municipal water department engineers need to work together in the districts, especially as there is no map of the network. However, in a further show of opposition, the engineers are often not prepared to reveal the location of the pipework to the consortium, who, in turn, consider this refusal a form of sabotage or blackmail.

Is this actually opposition, concern over the arrival of a private operator or a simple refusal to cooperate? This question applies not only to the water department engineers but also to the various public administrations (municipal and regional). With new urban policies being launched to eradicate slums, there is potential for rehousing programmes. These programmes are implemented by the Slum Rehabilitation Authority (SRA), a parastatal organization. Where these involve resettlement, cooperation between the SRA and the private water operator is vital for ensuring that the water network is extended to these new housing developments. However, the SRA has never been officially informed of the plans to delegate the water service.[11] The head of the SRA has worked at the local municipal council and, like all SRA staff, was only too aware of the contract with Orange City Water. However, this senior civil servant clearly feels that the SRA was sidelined from discussions and is thus now reluctant to accept this "privatization". This shows that, even though urban systems in India are poorly coordinated and highly fragmented, there are robust informal municipal employee networks in place through which information is effectively channelled.

As existing stakeholders often see the introduction of private operators as signalling a loss of authority and a shift in the balance of power, it may seem surprising that this opposition and informal flow of information were not taken into account when the joint venture was being set up. A prime strategic challenge here involves getting to grips with the twists and turns of a fragmented yet interconnected state to best ensure its cooperation. First, although the company was relatively surprised that nobody wished to work for them, Veolia's managers, in charge of the company's operational performance, took the view that this meant they were spared having to work with employees who were not motivated and used unacceptable practices.[12] Furthermore, they also thought they would have greater opportunity to recruit and train their own employees; however, in reality, this opportunity has been limited due to the lack of people in India with the relevant skills.[13] Second, the division of responsibilities between Veolia and VIL for engaging in discussions with the other departments involved in urban development in Nagpur appear to have been clearly defined. The role of VIL, which has skills and knowledge relevant to the local area, was to manage the relationship with all of Nagpur's public authorities. VIL's director was given the task of maintaining a good relationship with all public administrations; however, it appears he refused to do this while, at the same time, also categorically refusing to give Veolia's managers the go-ahead to manage this relationship in his place.[14]

This situation is partly due to the fact that the two consortium members' strategic objectives are poorly aligned. Veolia clearly selected VIL to ensure they had a partner who was well versed in the workings of the state within Nagpur. There were other more well-established and respected potential partners available; however, these were all unfamiliar with Nagpur. The close relationship between the head of VIL and the leader of the political party in power also played a key role in Veolia's choice of partner. This close relationship has already helped VIL win many other contracts (Menon, 2012). Thus, this arrangement is not based on the familiarity that Luo (2009) highlighted as being important, but instead requires the local company to act as broker. In contrast, one of this local company's objectives is to develop its activities within the water sector or other infrastructure-based sectors. VIL's involvement in this contract is intended as a stepping stone for responding to other calls for tender in the future. In other words, one of the joint-venture partners is seeking to establish a local presence, whereas the other is looking to distance itself from its local roots. This is illustrated by VIL's attendance at international conferences where it presents its experience of working in Nagpur without mentioning its partnership with Veolia.[15] VIL's distrust of the foreign staff sent over by Veolia to oversee the project's operational management should also be viewed through the prism of this objective. Indeed, it was a power struggle between VIL and Veolia that led to the departure of the head of Orange City Water, a foreign member of Veolia's staff, with VIL's preferred management recruitment strategy being to hire staff from the Asian labour market. As a result, a number of managerial posts remain vacant, undermining the relationship with the municipality (Anparthi, 2013). The success of the joint venture is thus partly dependent on who has control over human resources. However, the partners also have disparate strategies for achieving the aim of reducing costs while also building expertise, as illustrated by the conflict surrounding subcontracting.

Subcontracting under pressure: minimizing costs and developing technical skills

In India, the public sector has always used a range of subcontractors for carrying out construction work. These – often small – construction companies form part of a value chain in which traditional skills are preserved and which operate through patronage networks at the same time as evoking discussion on "good governance" (Hibou, 1998; Zérah, 2009). The selection of subcontractors has been the source of serious disagreement between the partners, again due to a lack of alignment in the strategies of the two companies. For Veolia, the most important selection criteria were technical expertise and the ability to meet the required deadlines and establish credible and transparent contractual relationships. By focusing on these criteria, Veolia thus expanded its list of subcontractors to include those outside of Nagpur and began working with a number of these companies. However, VIL did not approve of this approach and refused to authorize payment for the work carried out. There are two possible reasons for VIL's position. First, since the aim is to minimize costs, profit margins are likely to be small and the work poorly supervised and of low quality, which is an issue for all similar projects. The head of a company involved in an Indian–Israeli partnership working on a project in an area of Delhi confirms that the main bone of contention can often revolve around the strategy to reduce construction costs to increase profit margins, which are very small compared to other countries.[16] Sent over from abroad to manage the consortium, he explains that his role as the only foreign employee and head of the company is to ensure that quality standards are upheld. Second, VIL carries out its work in the region through a network of subcontractors, which are also instruments of a paternalism that helps circulate and allocate the revenue generated by the contracts. Work to establish a code of conduct setting out the relationship with subcontractors and employee working conditions, and which could have fostered discussion and joint decision-making, has been put on hold. There are mounting delays, the consortium's reputation among construction companies has been tarnished and its inability to improve working conditions could prove highly damaging at a time when discourse on social responsibility is being established.

Thus, in contrast to those authors who have shown that equal ownership reduces conflict through increased transparency and a more equal balance of power (Luo, 2009), for Nagpur, evidence points to the opposite being the case. As highlighted by Wu and Malaluan in their analysis of two '*concessionnaires*' involved in the privatization of the water sector in Manila:

> although a strategic alliance is necessitated by political, legal, financial and technical considerations, the potential for conflicts of interest among them cannot be underestimated. Bad corporate governance can quickly lead to internal conflicts that may bring out the worst in all involved.
>
> (Wu and Malaluan, 2008: 211)

In Manila, the fact that consortium members awarded outsourcing contracts to their respective subsidiaries became a source of conflict and limited their ability to react to the contract's loss of financial viability caused by devaluation of the currency and failure to adjust the tariffs. In Nagpur, having joint decision-making authority has created bottlenecks, as decisions are based on contradictory selection criteria: technical skills versus maintaining a local patronage network of subcontractors, which also requires the political cost of decisions to be taken into account.

Connecting the districts: the social and environmental challenges

The plumbers: the invisible drivers of access to the service

Since it accounts for only a very small amount of the cost compared to the rest of the network, the last link in the water service production chain, namely the household connection, is often overlooked. However, this tertiary network can prove to be a weak link at which leakages and illegal or substandard connections may be found. Covering this section of the network is a group of professionals that is often disregarded but clearly identified in the study by Rémi de Bercegol, and which plays an intermediary role: the plumbers. There are around 450 plumbers in Nagpur and 150 of these are licensed by the municipality, paying an annual licence fee. These are not municipal employees but they fulfil a role that the municipality is no longer able to assume as it has stopped recruiting. They install household connections, giving them a reliable stream of income, and provide advice and information to water service users.

However, municipal licensing and installing connections are linked to a hidden economy, the cost of which is subsequently borne by the customers:

> When Nagpur municipality knows that a private plumber has illegally installed a connection, it immediately has it disconnected. When an illegal connection is installed by a licensed plumber, the administration and the plumber make a deal that means the connection remains in place.[17]

The role of these licensed plumbers is thus central to customer relations: it is estimated that 30 per cent of all connections to the water supply network in Nagpur (and 60 to 80 per cent of connections in the slums) are illegal. The plumbers are thus the advanced arm of a malleable and complex local state, the boundaries of which are blurred, as reflected in the porous nature of the relationships between intermediaries, local elected officials and the bureaucracy (Benjamin, 2005). These plumbers have held large-scale and frequent anti-privatization demonstrations and protests since the contract was launched. Those elected officials supporting the demonstrations consider these plumbers to be an extension of the state, as the long-standing failure to recruit municipal staff has made it necessary to maintain permanent relationships with these subcontractors. The plumbers, meanwhile, are demonstrating because they fear losing their income and their jobs, both of which are dependent on their relationship with an ambivalent state. However, these patronage relationships between a malleable state and the plumbers, who have become well known for their use of illicit practices, are roundly condemned by the delegated management company as being part of a system that is riddled with inefficiencies and corruption, and which excludes the customers' needs.

The social aspect of access to the network

Indeed, the relationship between the customers and network managers is central to any distribution service. As noted by the manager of another delegated service in Delhi,[18] the plumbers' role is not confined to poor districts. Being responsible for connections, they also negotiate directly with customers from wealthy districts to install larger connections, which "destabilize" the network and complicate the long-term technical management of the water scheme. However, it is in the districts with no connections to the water network that the issue of inclusion is the most pressing. In Nagpur, the contract includes connecting and billing all customers, as well as managing the network of standpipes installed by the municipality. For all new connections, the operator has to review the connection requests before forwarding them to the municipality for approval. In practice, this creates uncertainty over whether it is possible to connect customers or not, which partly corresponds to the duality between notified districts (which may include slums) and non-notified districts.[19] De facto, certain districts may be deemed ineligible for connection campaigns, whereas others may receive services, despite being illegal, due to their political support. Mapping the various means of access to water services provides an illustration of this, as there is no direct correlation between the service level and legal status of poor districts; other factors linked to political patronage can also help ensure that a district gains access to water (Bommier *et al.*, 2012).

The inhabitants of districts without a water service therefore experience what Wolff and de-Shalit (2007) call "corrosive disadvantages": they are disadvantages in terms of access to a basic service such as water and dependent on local elected officials, who use their votes as a bargaining tool and thus take away their political autonomy. However, at the same time, as certain authors have shown, political brokering can also provide access to services for unauthorized districts (Benjamin, 2005). Thus, in terms of social responsibility, the challenge for the joint venture is to define ways of ensuring that all customers will ultimately be connected to the network. The decisions that need to be made to extend the network cannot be based on the technical criteria alone. While the municipal administration is officially responsible for approving the network extension plans, this approval is also informally shaped by political priorities and the influence of certain elected officials. To facilitate expansion of the water supply network, Orange City Water held extensive discussions with the municipality to introduce free connections, the idea being that the connection costs would be covered by the funding received through the JNNURM urban modernization scheme.

Once the expansion programme has been defined, Orange City Water develops a social engineering approach to be implemented by a specifically created team from within its sales department. This team's role includes communicating and liaising with the poor districts. There is a field coordinator for each area who explains what the company is planning to do, helps people complete their connection requests and handles everyday issues. Service points, or customer outreach centres, have also been set up in each district. The partnership between the two companies that make up Orange City Water appears to work well in relation to this approach: approval of the operational strategy went smoothly, which facilitated expansion plan discussions with the municipality.[20] However, although access to the water supply network is currently being extended and is supported by the national programme, the contract does not cover wastewater evacuation, which falls outside of Orange City Water's scope.

Beyond the scope of the contract: the environmental aspect

Nevertheless, Orange City Water is rightly concerned about sanitation, for two reasons. The first relates to the pressing need to improve the sanitation situation. The second pertains to Nagpur's location, namely within a region that significantly lacks resources. The economy in rural areas of the Vidarbha region has been in crisis for many years now and thousands of farmers have committed suicide as a result.[21] Population forecasts for the next 20 years indicate that daily water requirements will rise and the city of Nagpur is thus seeking to increase its production capacity; however, neither the impacts of this urban policy nor the ecological footprint of this development have been taken into account.

The situation within the region has attracted the attention of highly committed environmental activists with diverse influences. Ever since the launch of the new water supply projects, these activists have been challenging the decision to focus on the city area, as the additional extraction of water from the river could severely affect the lives of people in the rural areas downstream of the water treatment plant. Thus, one of these activists stated: "there may be a famine in the future; we have to fight for water in this project. Can we have water? We are willing to sacrifice ourselves".[22] Another campaigner, who describes himself as an eco-social activist, is critical of Western development models:

> We reject the IMF and World Bank influenced neo-colonialist agenda. These predatory policies, which don't care about the people, are directly responsible for the separation between urban and rural areas, for the opposition and protest movements, which lead to shameful crises, such as farmers' suicides.[23]

A third activist, influenced by Gandhi, highlights the impasse created by promoting industrialization without taking the sharing of available resources into account: "We have enough natural resources to meet the basic needs of all the country's citizens.... Problems arise from a lack of planning and implementation or poor resource use. That's what I think".[24] With the arrival of the private company, the heavy criticism of the Indian development model has thus had an impact locally. The development model's aim is to promote sustainable development, which should also include developing the linkages between access to water and sanitation. Although aware of this, the company, which has endeavoured to encourage the public authorities to address sanitation issues, is itself unable to do so within its current contractual framework. It does not have the legitimacy required; yet, at the same time, the defined contract scope is unable to protect the private initiative from criticism, which has been heightened by its arrival. This situation is significant due to the questions it raises about social responsibility when the service being provided is above all a public service.

Conclusion

The review of the Nagpur contract has invalidated an argument frequently put forward in literature on joint ventures, namely that equal ownership fosters compatible decision-making: it is not only the technical decisions (who to connect first, what type of meters to install), but also the way in which the joint venture defines its relationships with its essential intermediaries, its subcontractors, that are important. However, the relationship with these subcontractors cannot be viewed merely as a formal contractual relationship; it also involves other factors that draw upon the porous nature of borders and upon informal networks. Furthermore, the fact that the company remains highly dependent on public institutions makes it difficult to implement its stated goals, particularly with regard to connections. This is because it is up to the municipality to approve expansion plans and to lead broader discussions on environmental issues, which it is reluctant to do partly because anything outside its municipal boundaries is not its responsibility. The discourse on social and environmental responsibility thus seems to clash with the tension that exists between universal access goals and empirical reality. Indeed, actual field practices are dependent on an urban service provision economic policy and power relationships in which the inclusion of all is only secondary. It may also be the case that, during the course of the contract, the expectations placed on multinationals and their voluntary commitments could bring greater equity to local urban capitalism.

The Nagpur project indicates that there are several ways in which improvements could be made. First, some of the regulatory requirements, such as those pertaining to subcontracting, need to be improved and indicators included that take environmental issues into account. Second, there needs to be a more in-depth review of shared governance mechanisms that foster the involvement of elected officials, user representatives and non-governmental organizations, not only in the implementation of urban public service projects, but also in ensuring that the actions undertaken meet contractual objectives and even in analysing the social impacts of the contract scope. However, both municipalities and companies are often reluctant to introduce participatory processes.

Lastly, the review of the situation in Nagpur reveals the difficulties encountered by private–private partnerships when carrying out a public service role. The multinational company feels an overriding sense of frustration, despite the efforts made to improve its understanding of the local environment. For the local partner, the multinational's focus on technical skills, upon which it has built its reputation, has threatened to reduce the amount the local company earns from infrastructure projects. Rather than creating long-term expertise and urban service industry leaders, national network funding policies thus seem to have instead increased the number of entrepreneurs looking for opportunities to deliver projects at low cost. Finally, having been reviewed by international institutions, the public authorities, the water customers, their representatives and various local and national associations, the establishment of these contracts also gives rise to further issues. These include finding the resources to promote new coordination mechanisms while safeguarding public interests, dealing with issues relating to cities' ecological footprints and examining the tools available for overcoming the lack of service provision that has been common in large parts of cities for decades. As such, these partnerships have opened up debate on the way in which technical systems have been installed and managed to date.

Notes

1 The project implemented in three small towns in the south of India won the National Urban Water Award in 2009. See Kumpel and Nelson (2013) and Water and Sanitation Program (2010).
2 Communication from the Commission to the European Parliament, the Council, the European Economic and Social Committee and the Committee of the Regions, A Renewed EU Strategy 2011–2014 for Corporate Social Responsibility, 681, 25 October 2011, p. 7.
3 See Article 135 on CSR, p. 80. Available at: www.mca.gov.in/ministry/pdf/companiesact2013.pdf) and, for a critical analysis by a senior civil servant, see Arun Maira (2013).
4 Recognized as a fundamental human right by the General Assembly of the United Nations in July 2010.
5 These issues have been clearly highlighted in examples from the clothing and sporting equipment manufacturing industry and oil sector. The best-known example in the water sector is that of Cochabamba in Bolivia.
6 The Municipal Commissioner manages the municipality and has greater authority than the mayor. In large cities, this position is held by a central government civil servant of the Indian Administrative Service.
7 Following a number of corruption cases, this president left his post and his political influence was much reduced, but he is now part of the BJP government led by Narendra Modi as the Minister for Urban Development.
8 Interview conducted on 29 November 2013 by Marie-Hélène Zérah.
9 This implementation period was reduced to two years, raising numerous operational issues.
10 In Delhi, municipal employees also refused to transfer to private companies.
11 Interview with the head of the SRA in August 2011.
12 These practices were reviewed by Karen Coelho as part of her ethnography of water engineers in Chennai. She clearly describes the way in which rules are manipulated and circumvented using practices that could be considered unacceptable, but which are often vital to ensuring the smooth running of the service; see Coelho (2006).
13 This situation was confirmed by the head of Suez in India. In addition, Veolia and Suez are competing against each other to recruit water sector engineers, both making job offers to employees of Jusco (a subsidiary of Tata, which has withdrawn from the sector following the failure of a contract it was awarded in a city in the south of India).
14 Interview, September 2012.
15 See the VIL website: www.vilindia.com/.
16 The Delhi Water Board (a public institution of Delhi's government) signed three management contracts with private consortiums in order to build and manage the network for 12 years in three districts of the city. The consortiums are partnerships between: Suez and SPML Infra Limited; SPML Infra Limited, Tahal and Hagihon; Veolia and SREI Infrastructure Limited.
17 Interview conducted in August 2011 with an unlicensed plumber in a district of Nagpur; author's translation, in Bommier *et al.* (2012: 136).

18 Interview conducted on 23 October 2013.

19 In India, each regional state has graded notification criteria for unauthorized districts. In principle, unauthorized districts, which are not necessarily slums, do not have the right to receive urban services. Notified districts, which may be slums, are able to access urban services. There are also some districts in a halfway state, whose status is somewhere between "illegality" and legitimacy.

20 According to the head of Veolia India, the rate at which new connections are being installed is higher than on any of Veolia's other contracts.

21 The agrarian crisis of the past few years has led to waves of suicides among farmers in numerous regions of India. In Vidarbha, the situation is particularly worrying and has prompted the government to introduce a rehabilitation package; see S.S. Kalamkar and Sangeeta Shroff, "Impact of Rehabilitation Package in Suicide-prone Districts of Vidarbha", *Economic and Political Weekly*, 5(46), 2011, pp. 10–13. The India Social Institute documented the situation in 2012 and estimated that there have been over 2,000 suicides in the region since 2001. Available at: www.isidelhi.org.in/hrnews/HR_THEMATIC_ISSUES/Agriculture/Agriculture-2012.pdf. For information from farmers' organizations see Vidarbha Jan Andolan Samiti's website: http://andolan.blogspot.in/.

22 Authors' translation (Bommier *et al.*, 2012: 101).

23 Ibid., p. 103.

24 Ibid., p. 107.

Part III

When small water systems meet the large technical system

7 A local public service

The action of small-scale private water providers to supply peripheral neighbourhoods in three African cities

Sarah Botton and Aymeric Blanc

By 2020, Sub-Saharan African cities will be home to half of the continent's population, including 50 per cent in informal settlements. In a context of scarce public resources and where users cannot bear the cost of basic services alone, the water service is the most symbolic sector for this gap between urban growth and the provision of essential services. This is because it is widely subject to an economic, technical and environmental rationale (scope of infrastructure, technical system, cost of service, etc.), and comes up against strong health, social and political expectations concerning access to water.

There is such a high infrastructure deficit in large African cities that water distribution systems generally only cover part of the urban fabric, leaving vast areas unserved (Collignon, 2007). The informal settlements that have sprung up over the past 30 years experience a host of difficulties: technical (narrow streets, flood-prone land, rugged topography), commercial (identification of plots, addressing), legal (clients without land titles, neighbourhoods that are not in the land registry), political (slum clearance projects often block supply projects) and institutional (it is not always clear who decides whether or not to supply these areas with services) (Collignon and Botton, 2014). Consequently, the residents of these neighbourhoods buy from informal operators who work in gap areas not covered by the main operator.

For a long time, stakeholders in the water sector, especially public institutions, focused on economic considerations concerning *the* optimal technical system (a centralized network that necessarily allows economies of scale), *the* appropriate management method (water authority, delegation, semi-public company), *the* most effective delegation method (concession, affermage, management contract) and *the* right level of service (national, regional, local). They are gradually beginning to put the strength of the models that are implemented into perspective in order to observe their practices. Certain experiences, which are often heterodox with regard to theories that have shaped the sector for over 30 years, have consequently attracted attention, such as the small-scale private water providers who work in the water sector. These small-scale systems provide quality of service that is sometimes comparable to the provision of the main operator. This calls into question the main principles of the economic theory of the natural monopoly of water services and of economies of scale, which are the cornerstones of the concept of public services for water.

The study of small-scale private water providers necessarily leads to focusing on their position at the crossroads of the professional world of water distribution on the one hand, and informal settlements, with their populations, associations and political representatives, on the other hand. It shows that they appear "to play an essential role as an intermediary and interpreter between different and even contradictory rationales for action" (Crozier and Friedberg, 1977: 86).[1] In addition to a normative assessment of the services they provide ("pioneers or predators?") (Hailu *et al.*, 2011), it is necessary to understand in what way cooperation and contractual arrangements between the main operator and small-scale providers impacts upon practices in the water sector.

For some 20 years now, the literature on these small-scale providers has given no explicit definition of their function. The World Bank and Water and Sanitation Program (Snell, 1998), Asian Development Bank (Conan, 2003) and Agence française de développement (Blanc *et al.*, 2009; Botton and Blanc, 2010) all refer to the great diversity of forms of small-scale providers.[2] However, they are generally defined by their "informal", "spontaneous" and "independent" nature;[3] the urban areas where it is difficult to serve clients from low-income households with a conventional network; the high prices for services provided by resellers and the resulting unfairness, which feeds into the notion of the "poverty penalty" (Hailu *et al.*, 2011); the flexibility of their services and the complementarity they provide with the conventional network (Solo, 2003; McGranahan and Satterthwaite, 2006). While their service is less sophisticated than the main operator's service and is sometimes "cobbled together", it is nevertheless functional and, especially, accepted locally (Valfrey-Viser *et al.*, 2006).

On the contrary, in this chapter it is the "regular", "encouraged" and "coordinated" nature with the main operator that will be studied, on the basis of a comparative analysis of three large African cities (see Table 7.1): Maputo (Mozambique), Ouagadougou (Burkina Faso) and Kisumu (Kenya), which have recently implemented pilot projects to delegate water services to small-scale water providers to supply outlying neighbourhoods (with financial support from Agence française de développement). These three cases have a diversity of configurations in terms of the status of the main operator, institutional functioning, and the accessibility conditions to water resources. The aim is to examine the extent to which cooperation mechanisms among official services are able to handle the technical and commercial management of water services in areas that are often deemed "physically or financially unattractive" (Ahlers *et al.*, 2013).

The gradual emergence and legitimization of the figure of the small-scale water provider

In the 1990s and 2000s, with the proliferation of schemes to delegate water services to international operators in large developing cities, there was the

Table 7.1 Summary of the three cases presented

	Maputo	*Ouagadougou*	*Kisumu*
Population	2 million	2.3 million (+7% a year)	500,000
Population not connected	1 million	200,000 (10 unconnected, informal neighbourhoods)	320,000 (including 250,000 in unconnected informal settlements)
Population concerned by the pilot project	100,000	86,765	60,000
Organizing authority	Ministry of Public Works and Housing (MOPH)	ONEA under the supervision of the Ministry of Water	Government
Main system operator	AdeM → FIPAG (2011)	ONEA	Kiwasco (2001)
No. of connections for small informal systems	60,000: 30% of the city	–	–
Operators of small informal systems	450/900 SSWPs	–	–
Operators of small systems under PSD – Pilot projects	16 SSWPs under PSD with Fipag	5 delegatees under contract with ONEA (5 neighbourhoods)	5 Master Operators under contract with Kiwasko (Nyalenda neighbourhood)
Selection method, duration	Bid invitation, 5 years	Bid invitation, 1 to 5 years	Official advertisement and interviews; 3 contracts with NGOs, 2 with local entrepreneurs
Population connected	12,000 connections obj.: increase from 40% to 70% (in IC) by 2015	3,640 connections (i.e. approximately 32,000 h)	From 19 to 73 connections depending on the networks (36 people per connection) 15,000 h max (5 × 3,500 h)
Length of network installed	173 km	60 km + 65 standpipes	Five 600 m pipelines
Volume produced	+7,500 m^3/day (target)	Bulk purchase	Bulk purchase
Volume distributed	–	3,600 m^3/day	–
Origin of drinking water	Boreholes	ONEA, GST	Kiwasco, GST
DW price	Aligned with AdeM tariff (so below informal SSWP tariffs excluding PSD)	ONEA national tariff schedule	Regulated < GST price

promise of waging "the real struggle for water" (Mestrallet, 2001) in order to allow access for all. Furthermore, the Millennium Development Goals (MDGs) declaration was signed at the UN in 2000, whereby the governments of 193 countries committed to quantified targets. This period allowed the controversial experimentation of pro-poor mechanisms in informal settlements (Botton, 2007). However, on the whole, privatized companies did not come up with the economic and social miracle that had initially been promised (Stiglitz, 2002). Following a few very symbolic failures of large-scale water concessions (Cochabamba, Buenos Aires, Jakarta, La Paz, etc.), a new change of course led to the reintroduction of the principle of the "local policy" for water and the return of decentralized public authorities to centre stage (the turning point came at the World Water Forum in Mexico City in 2006). Donors proposed a "reviewed" private sector participation, promoting "local actors", including small-scale water providers. Consequently, there was a shift from "access to drinking water services for all" to "improved access for the greatest possible number of people".[4] While small-scale water providers had always existed in the informal sector, they gradually appeared as one of the "potential" solutions to the political problem of access to drinking water.[5]

Since then, the legitimization of small-scale private water providers has been put on the agenda, in particular for pilot projects financed by certain donors or public authorities. From the first World Bank publications on this topic in the late 1990s, small private providers appeared as "intermediaries" for the main operators who could not manage to achieve their service objectives, as well as the "innovative solutions" so much sought after by the MDGs (Cavé and Blanc, 2011). It was thought that they would improve the service in areas to which official operators did not extend their services, while creating significant financial leverage for the sector (through their own financing). Consequently, these mini-urban networks, under contractual agreement with the main operator to provide a technical and commercial relay in informal settlements, are seen as a system that is likely to "transcend the obsolete distinction between public service (utility), which is characteristic of the network management, and market service (commodity), which is specific to 'informal' supply networks" (Jaglin, 2005: 149).

Comparison of the projects of small-scale providers in three large African cities

Maputo-Matola: initiatives formalized by contract

The water sector in Mozambique was subject to major reforms in the 1990s, with a decentralization movement and the introduction of private sector participation. The delegated management framework introduced is, first, based on an asset management company (Asset Holding and Investment Fund for water supply: FIPAG), which is under the supervision of the Ministry of Public Works and Housing, the owner of the water sector infrastructure for the country's main cities; and, second, on an independent regulator (Water Regulation Council), which is responsible for delegated management contracts (Ginisty, 2009; Matsinhe *et al.*, 2008a, 2008b; Chaponnière, 2008). It is an original example of low-cost household connections for populations who are underserved or unserved by the main network of the operator Àguas de Moçambique (AdeM). A number of small-scale private water providers have spontaneously developed in the city, while some contractual arrangements have been established.

In 1999, in the metropolitan area formed by Maputo, Matola and Boane (around two million inhabitants), the water service was delegated to the international operator AdeM *via* a 15-year affermage contract signed with FIPAG. The service is particularly deteriorated in peri-urban areas in the northern half of the city that are not covered by the network. The latter has high levels of non-revenue water and uses surface water that is treated in a plant far from the city centre. Informal suppliers are motivated by unsatisfied demand for water and, since the late 1980s, have invested in creating small supply systems using local groundwater. The way in which they have sprung up is striking: it is estimated that there are now some 900 private operators in Greater Maputo with over 60,000 domestic connections, serving about 30 per cent of the city.

In May 2008, the Minister for Infrastructure declared that these small-scale private providers "should no longer be countered, but rather formalized". Under the impetus of donors, ideas for formal service delegations sprang up with them in the areas that were not yet served. Donors are particularly present in Mozambique, where development assistance accounts for approximately half the state budget. They finance the rehabilitation of water supply infrastructure, drinking water production and the extension of the service to peri-urban areas. For example, a project financed by AFD for water supply in peri-urban neighbourhoods comprised the construction of 16 new, independent production systems, with a five-year delegation of operation to small-scale providers selected for their technical and financial capacities. This project aimed to increase the water production capacity by 7,500 m^3 a day for a supply to some 100,000 inhabitants. Furthermore, the project planned to densify existing small networks and subsidize individual connections for the poorest populations thanks to an Output-based Aid mechanism.

All the contracts, which have been signed for a five-year term, provide for the definition of a non-exclusive service area, setting out the obligations and rights of the delegatee, as in a "concession" affermage contract. The small-scale provider is responsible for the development and maintenance of secondary and tertiary networks (depending on the agreed technical specifications) and for supplies for the installation of the electromechanical equipment. It is required to supply high-quality water and service, establish an appropriate client management, conduct regular reporting on its activities, operate the system at its own expense and risks (energy, salary, chlorine, tooling and spare parts, etc.), and pay rent for the provision of the facilities (approximately US$0.1/$m^3$). It is remunerated, first, on the basis of water sales, and second, on the basis of subsidies (which especially contribute to approximately halving the amount paid by new clients for an individual connection). Water sales are made according to the tariff schedule: these sales include both the volumes sold at standpipes and the sales from individual connections (calculation on the basis of a fixed part of 1 m^3 and a variable part of 18 MT/m^3, combined with a price revision formula). As for the delegator (FIPAG), its main undertakings are to maintain an adequate production capacity and pay the subsidies owed to the delegatee.

While 173 km of secondary networks and 12,000 connections have been installed, the entire system has been located in neighbourhoods far from the extension areas planned for the main network. Furthermore, despite communication efforts, the authorities in charge of the project came up against great mistrust and hostility on the part of the small-scale providers: only a few accepted the switch from the role of being an owner of infrastructure to that of a tenant having to pay a fee to use it. Others saw these new investments as a threat to their activities, whereas they would have preferred access to financing.[6]

Ouagadougou: extension of main operator's action to informal neighbourhoods

The experience of Ouagadougou, the capital of Burkina Faso, provides an example of an operator seeking intermediaries who could operate under its control and manage the service in outlying neighbourhoods. The National Water and Sanitation Office (ONEA) has strong support from the Government of Burkina Faso. This state-owned company, with a capital of €4.7 million, has been in charge of water services in Burkina Faso since 1994. It is under the technical supervision of the Ministry of Water and operates nationwide by serving the 47 towns in Burkina Faso that have over 10,000 inhabitants.

The city of Ouagadougou is characterized by strong and uncontrolled urban development: 2.3 million inhabitants and an annual growth of about 7 per cent. In 2012, while the water access rate in urban areas stood at 84 per cent, it was assessed to be 98 per cent in Ouagadougou in ONEA's area of responsibility. However, it does not include populations living in informal areas (ten or more neighbourhoods with a total of some 200,000 inhabitants). Furthermore, the construction of the Ziga dam in 2005 resolved the problem of the city's water shortage (Hydroconseil, 2013).

ONEA, with AFD support, has developed a project to serve informal neighbourhoods by establishing public service delegation contracts with private operators. The aim is to increase the water access rate for some 100,000 people (50 per cent of informal neighbourhoods). The project plans to deploy networks in five of Ouagadougou's neighbourhoods, with the technical and commercial management entrusted to local private operators under contract with ONEA. There are two possible supply methods: by private connection or standpipe. These delegations are the result of a bid invitation for which five lots had been defined, combined with contracts for one to five years. The five delegatees were selected (out of 20 or so candidates) on the basis of their financial bids concerning the purchase price for water from ONEA, and the subsidy requested per private connection. The missions expected of the delegatee are to handle the technical operation of the network, the programming of future investments and the commercial management of customers.

The delegatees have varying profiles. None of them have previous experience in the sector, but two have an activity related to the water sector (design office and construction of tanks and metal parts for water towers). The others are small trade structures (import/export). All the delegatees are required to apply ONEA's tariff for connections and water sales for both individual connections and standpipes. Target results are set out in the contract. The primary contractual concern is the number of individual connections and, second, the volume of bulk water purchased annually from ONEA. If these targets are not achieved, ONEA can impose penalties on the operator.

The first years of activity increased the access rate for populations in the five informal neighbourhoods to 90 per cent, with 60 km of networks and 65 standpipes. While at the end of September 2010 the delegatees had already made 712 private connections, it should be noted that in some cases the results targets have partially been achieved (20 per cent), whereas in others they have been largely exceeded (470 per cent), thereby suggesting that they had been poorly assessed beforehand. The number of individual connections is tending to increase as a result of the reduction in the connection fee, which fell from FCFA 50,000 to FCFA 30,500 at the end of 2010 – it dropped to FCFA 3,640 two years later. The objective for annual water purchases from ONEA (which was poorly assessed beforehand) has been achieved by all the delegates; some have even exceeded it by almost six times. Consequently, some 690,000 m^3 were distributed by the delegatees in 2010 (1.3 million m^3 in 2012), which corresponds to an average consumption of over 21 litres per person per day.

The operators provide a local public service mission, which ONEA could not provide (informal area, close local relationship) and the mayors are satisfied with the experience. Furthermore, the delegatees appear to have a sound technical management of the networks (the network performance is estimated to be over 93 per cent, with all the precautions required by this type of figure), even if most of them do not have a technical profile in their permanent teams. Consequently, the delegatee's competence as a manager-accountant would appear to go further than a competence for water. Four delegatees have achieved positive results, which increased during the first year of operation, in particular due to equipment procurement strategies. Some delegatees would even be willing to extend networks using their own resources. In addition, there has been a significant increase in the volumes consumed, with a constant or even lower cost per cubic metre of water, combined with a gain in terms of convenience (reduction in distances for carrying water, development of latrines and showers, job creation, etc.). However, the delegatees still need support for certain issues, particularly for the ownership of facilities, the guarantee period for facilities, the contract review in the event of housing schemes in neighbourhoods, technical management, especially the identification of losses, the control of works by ONEA, etc.

Nyalenda (Kisumu): the professionalization of neighbourhood associations

The experience of Kisumu provides an example of water services subject to technical and commercial difficulties – 36 per cent coverage of the population, technical and commercial losses of 67 per cent – and an operator who needs to supply a city with 50 per cent of informal settlements (against about 10 per cent in Ouagadougou). Despite the apparent abundance of its surface water, Kenya officially ranks as a country suffering from "water stress" (less than 1,000 m^3 per capita per year). Kisumu is located on the shores of Lake Victoria and is the country's third largest city with some 500,000 inhabitants. Almost 50 per cent of its population lives in unconnected informal settlements, while a large part of the network supplies poor-quality water, which is sold at a high price. The government launched a major reform of the sector by adopting a Water Act in 2002, with assistance from donors who provide financial support and technical expertise. Consequently, in 2005, the municipal company was able to rehabilitate and extend water and sanitation infrastructure. A project initiated by French Cooperation, in partnership with the World Bank's Water and Sanitation Program (WSP), also comprises an original component for informal settlements.

The specificity of this project in the Nyalenda neighbourhood (60,000 inhabitants) lies in the fact that it organizes the delegation of the water service to local operators. The aim is to develop subsystems managed in a community-based and participatory manner by a delegatee from the neighbourhood – an NGO, association, private individual or commercial company – (called the Master Operator), under contract with the water company KIWASCO. The company already had "privileged access" in the neighbourhood via a small, flexible and very shallow underground pipeline network, some of which was equipped with meters and gave rise to a (very low) billing. The bulk of this network was made up of illegal connections, leading to leaks and fraud, and consequently to non-revenue water for the operator. The establishment of a delegation scheme was therefore initially devised as a way of formalizing not only commercial relations with users in the neighbourhood, but also the technical distribution system (Castro, 2009): five pipelines (600 m pipes) were installed, with the aim of each supplying about 12,000 people. Each pipeline successively gave rise to a delegation. The process was conducted in a selective manner: following an official advertisement, the operator and Nyalenda Water and Sanitation Committee met, interviewed and selected the most qualified neighbourhood associations or small-scale local private entrepreneurs.

The small-scale provider is in charge of the technical management (minor maintenance) and commercial management (billing, collection) for the service. It is authorized to extend the network, but purchases bulk water from the main operator at a flat rate (KES 25, i.e. US\$0.35/m^3) and subsequently resells it to users at a regulated tariff (lower than the official tariff schedule). There are three possible water distribution methods in Nyalenda: (1) residents can opt for an individual subscription with a domestic connection and a meter; (2) there is the possibility of sharing a water point (for tenants), in which case the owner pays an overall bill on the basis of the consumption measured for each tenant; (3) the most common practice is when users buy water at a kiosk, a small independent retail outlet where the water resale tariff is controlled by a small-scale provider.

Consequently, the delegatee's main activity is the commercial management of the service, and the development of marketing tailored to the segment of poor clients in the Nyalenda neighbourhood. The daily door-to-door prospecting makes it possible to discuss health issues related to the service and the commercial terms being offered, and to negotiate to some extent with clients (e.g. payment by instalments). The main challenge for delegatees is, therefore, first, to sensitize the population of the neighbourhood to the benefits of a public water service and, second, to protect themselves against fraud and illegal water traffic.

In 2009, five delegatees had signed contracts with the municipal operator and some of them served up to 3,500 people. This delegation scheme provides the residents of the Nyalenda neighbourhood with closer access, which is up to ten times cheaper and of better quality. It also provides a strong incentive to reduce losses and increases the revenues of the main operator. The activity related to individual subscriptions is not very profitable for the delegatee (gross margin estimated at less than 10 per cent), in contrast to the activity of kiosks for which it is estimated that the gross margin ranges between 40 and 80 per cent, depending on rainfall (residents collect rainwater).[7] This result no doubt explains why delegatees have little incentive to finance the extension of the network.

Multi-speed public service?

Consequently, urban splintering places a strain on the practices of public operators (Graham and Marvin, 2001): while the public service must be universal, the users of this service occupy differentiated spaces. Recourse to delegatee small-scale providers in peri-urban neighbourhoods is conceived of as an extension of the main operator's action in areas in which it does not have socio-cultural entry points. The latter plays the role of "wholesaler", maintaining de facto a central place in the system (financing and construction of infrastructure, water production and transport), but it entrusts the small-scale providers with the complex mission of the customer interface (distribution, commercial management). Two questions are therefore raised. To what extent does the action of these delegatees constitute an extension of the public service provided in the central city? If economic theories justify the uniqueness of the operator in a relevant territory due to the existence of economies of scale, does the action of small-scale providers, in their role as "retailers", call into question the economic theories of the natural monopoly?

Out of everything that is expected of a public service, it is certainly the issue of service quality that is the most critical question for delegating operators. It is mainly based on three criteria: the bacteriological quality of the water distributed, the pressure, and the continuity of the service. As a result, the service quality provided by the delegatee small-scale providers is structurally related to the technical system that is implemented. In the case of Ouagadougou and Kisumu, where the delegatees, who are both "major clients" for operators and "retailers" for populations, distribute and sell water in neighbourhoods, it is mainly dependent on the service quality of the main operator, which produces and chlorinates the water itself. Moreover, in Ouagadougou, the evaluations have highlighted: a water quality similar to water distributed in developed areas, a pressure that is slightly lower, due to the "end of the network" effect and the fact that there are no water booster pumps, and, finally, in terms of service continuity, that the supply is admittedly sometimes cut off, but this is always attributable to the main operator. In Maputo, by contrast, as the delegatees produce and treat the distributed water themselves, they are much more responsible for service quality.

The second fundamental principle for the public service is equality between users, particularly in terms of tariffs. Out of the three cases presented, Ouagadougou is certainly the one that best embodies this pillar of the public service, with the obligation for delegatees to apply the national tariff schedule to users in informal neighbourhoods. In Maputo, the initial principle was supposed to be that of regulating the tariff by the market (selection criteria based on the lowest bidder in bid invitations). The regulator finally agreed to bring small-scale providers into the regulation and aligned the tariff for distributed water with the one applied by the operator AdM (Águas de Moçambique/Águas da Região de Maputo) (in fact lower than the many neighbouring informal operators). In Nyalenda, however, it is the principle of differentiation of service that is applied and therefore positive discrimination which has been chosen, with the tariffs offered to residents in the Nyalenda neighbourhood being lower than those applied in the centre of Kisumu.

Finally, the principle of mutability (also called adaptation) raises the question of the temporality and status of the action of small-scale providers: to rely on intermediaries who are certainly professional, but temporary, while extending the operator's service area, or acquire specific skills to work in informal settlements? Or permanently establish public service partners? Whatever the scheme implemented, the principle of mutability is present in the very spirit of contracts:

technical specifications, alignment of tariffs, short-term contracts, etc. are all elements that allow a potential reintegration of the infrastructure and services of small-scale providers under the umbrella of the main operator.

In addition to these three fundamentals of public service, there are the more recent principles of transparency on the one hand, and participation on the other hand. In terms of transparency, the projects in Maputo and Ouagadougou have shown that efforts have been made to allow bids to be advertised and explain selection criteria during the bid invitation phase. In this respect, the case of Kenya is less convincing. Similarly, the reporting obligations and implementation of a regular control of activities are part of this same concern. As for participation, while the social marketing activity is crucial for the small-scale providers responsible for sensitizing populations in informal settlements in order to get them connected to the system, the fact remains that at this stage the service on offer is neither devised nor co-managed with users. This is one of the critical points addressed by WSP in its analysis of the Kenyan project (Castro, 2009).

Conclusion

John Stuart Mill was far from Maputo when he laid the groundwork for the economic theory of the natural monopoly. Yet despite a telling antithesis embodied by the proliferation of independent networks (sometimes in the same territory!), the concentration of the sector currently underway would tend to prove that Mill was correct. In addition to the justification of economies of scale and the need to regulate natural monopolies, small-scale private providers have – at least for a time – shaken up the certainties of economists. Consequently, it would perhaps be possible to propose a single public service led by several operators, with a coordinated structure whereby the core network is managed by one party and the ends of the network managed by others. While the uniqueness of the operator would appear to be called into question by the three African cases presented, it does, however, seem that there are not sufficient elements to justify a possible "interruption" of the public water service: the extension (desired) of the water services of the central city would appear to operate, for better or for worse, with a uniqueness of service provided by a whole host of operators.

Questioning economic theories of the natural monopoly does, ultimately, involve an analysis of the power relations and scope of action at work with the intermediary mission of delegatee small-scale providers. These small-scale providers are caught up in a triangular framework between the main operator and the residents of informal settlements. Their relations with each stakeholder fluctuate, with varying degrees of autonomy and dependence, service provision and representation, commercial relations and cooperation relationships. They are sometimes ambassadors for poor populations, and sometimes the "small helpers" of water operators. They have a difficult role to play, whereby the public service for drinking water must above all remain a permanently profitable activity so that it continues to be worth the effort.

In certain cases, their role could be assimilated to subcontracting, as they would largely appear to be in a position of "providers of social engineering" on behalf of the main operators. In this case, the debate over the natural monopoly is closed, as the uniqueness of the operator and continuity of the networks and public service mission are consequently met. However, the establishment of "public service delegation"-type contracts positions them more, at least in theory, towards the responsibility for their service, as their remuneration is mainly related to the results of the service operation – the very basis of the notion of public service delegation. Consequently, questioning the theories of the natural monopoly means testing the level of coherence and homogeneity between the main service provision and the services of small-scale providers. If the predictions of economists turn out to be true, it would make sense for the small-scale private providers to disappear, to the benefit of a technical and commercial extension of the "official" water service in peri-urban neighbourhoods. If, by contrast, and as the evaluations of pilot projects would seem to suggest, small-scale providers appear to be an effective intermediary to provide both a better service at a lower cost and an activity that is profitable over the long term, it is consequently necessary to reconsider the idea of the natural monopoly in the local management of public water services.

Notes

1 The sociology of organizations refers to these operators as "intersecting marginal operators", i.e. drivers of innovation who interact with several systems.
2 The three standard figures described by Suzanne Snell (1998) are: partners of water utilities, vendors and resellers, and pioneers of small piped networks.
3 The adjective "independent" sometimes refers to the water source, to the non-exclusive and non-regulated nature of the activity, or to the absence of public subsidies; see Cavé & Blanc (2011). Similarly, a confusion is often made between the size of SSWPs (small/large) and their status (formal/informal). While it is true that informal actors often operate on a small scale, this does not mean that all micro-enterprises are informal.
4 See in particular the Joint Monitoring Programme between WHO and UNICEF set up in 1990 to monitor progress in global rates of access to water: www.wssinfo.org/definitions-methods/watsan-categories/.
5 Small-scale water providers include very different types of activities, ranging from various forms of reselling (fixed resellers, such as standpipe operators, mobile resellers of retail water in jerrycans, bottles, packs, ice cubes, etc., door to door: rickshaws, carts, tanker trucks, domestic resale, etc.), to more elaborate production and distribution activities (small-scale entrepreneurs, borehole and mini-network operators), and including community forms of service management (water users' associations, etc.).
6 However, the component of the FIPAG project, which aimed to subsidize SSWPs which develop new connections in their existing systems, was abandoned following unsuccessful attempts due to the absence of pre-financing for subsidized connections, the limit of the production capacity of SSWPs, or their fear that clients who could not afford to pay the full price for a connection would create additional collection problems.
7 According to the AFD team in Kenya.

8 Water cooperatives in La Paz and El Alto, Bolivia

A complementary system

Franck Poupeau and Sébastien Hardy

Introduction: research context, state of the art and hypotheses

In terms of urbanization in the countries of the "South", providing access to drinking water is a recurring problem. As the approach to the development of water distribution and sanitation systems based on the European model was implemented in Latin America, beginning in the twentieth century, the growth of cities, once associated with rural migration and recently with endogenous factors (Dureau *et al.*, 2006), has become out of synch not only with local water resources but also with technical capacities for extending elementary services, particularly water and sanitation services. But while political and, frequently, scientific debate focused on the status – public or private – of the operators concerned (Bakker, 2010; Bauer, 2004; Swyngedouw, 2004, Franco *et al.*, 2013), the technological pertinence and social effects of the universal distribution model – referred to as the "big system" – were rarely examined.

The "big system" is an organizational approach (either public or private) that continuously produces large quantities of drinking water distributed to consumers via a network of pipes in the hands of an operator that manages the entire municipal water cycle, from the abstraction of the untreated resource to delivery in the form of potable water. It constitutes a model in the sense that a homogeneous service providing drinking water to a large number of inhabitants requires a high level of technical and commercial skills (Mayntz and Hughes, 1988; Tarr and Dupuy, 1988; Lorrain, 2009; Lane, 2012). This is the model that is currently being used in the agglomeration of the cities of La Paz, the seat of government of the Plurinational State of Bolivia, and El Alto, the working-class periphery that, located on the *altiplano* (high plateau), overlooks its older neighbour. The "big system" was the object of a concession contract signed by a major international group specializing in the water sector (Suez). However, in 2007, largely for reasons of political symbolism, the Bolivian government terminated the contract and the concession was returned to the public sector.

However, the "big system" does not provide the only way of accessing water in the agglomeration of La Paz/El Alto. Many families procure potable water by other means: they get their water not only from networks managed by small cooperatives, but also in carboys. We will refer to these alternatives as "small systems". The process of local delegation of the big system has hidden the existence of other forms of provision likely to encourage differentiated social uses, which may be large and influential. Thus the distinction between "big" and "small" systems refers less to their importance in terms of water supply than to distinctive forms of management: while a big system is generally controlled by a single operator on a determined territory (Bakker, 2003), a small system is managed by the inhabitants of an area where the natural resource is transformed into a service for the community (Jaglin, 2003a).

The objective of this chapter is to analyse how these forms of provision function, to examine how and why other models of water procurement exist in the agglomeration side by side with the big system, and to study the ways in which these systems potentially complement one another. While there is an important literature on urban water supply and the fragmentation of the large technical systems (Graham and Marvin, 2001; Bakker, 2003), the literature on water has produced very little research on the coexistence of the two systems. Small systems are generally considered informal water supplies and not as coherent forms of organization (Misra, 2014), meanwhile academic literature generally analyses cooperatives as a form of economic organization in a business perspective (Young, 2002; Mayhew, 2008), to evaluate their performance indicators (Ruiz-Meyer and van Ginneken, 2006) or as an alternative to privatization (Hall, 2000), but not as a utility orientated towards other socio-political functions.

Moreover, the research presented in this chapter provides an original perspective on the social conditions of the possibility for collective self-organization of water uses, a concept suggested by Elinor Ostrom (1990, 2009) in her critique of the notion of the "tragedy of the commons" developed by Garett Hardin (1968). According to Hardin, collective ownership would lead to users exhausting resources, with everyone wanting to take advantage of the situation and serve their own interests: this would justify promoting individual property rights. For Ostrom, on the contrary, depletion of natural resources does not entirely account for the "ecological crisis". This crisis may also be explained in reference to the modes of "governance" of the commons and, thus, to the capacities of auto-organization in the process of allocating the resource and monitoring the uses to which it is placed. The Bolivian case is of particular interest in terms of sociological perspectives upon technical systems, not only because of the country's "water wars" (Cochabamba, 2000; La Paz–El Alto, 2007), which raised questions about the internationalization of private sector management of resources and services, but also because it highlights the way in which the emergence of new social uses transforms these technical systems and encourages the development of alternative modes of water management.

The survey described in this chapter is based in several sources of information: geographic information system (GIS) analyses, and interviews carried out by means of questionnaires and observations.[1] The database on risk management encompasses the public network (elements of the big system: springs, treatment plants, water transports pipes, wells, etc.) and resources sometimes very different from those generally used in urban contexts. For example, the wells used by private firms, including drinks companies, small cooperatives and companies using tank trucks, can, in certain circumstances, replace the big system. The information collected was homogenized as research progressed, and tables of data associated with specific geographical areas were developed. Based on these data, various maps were elaborated to support the analyses. For example, defining the percentage of households connected to the big system makes it possible to create spatial representations of alternative systems in certain areas of the agglomeration.

Some of the data were gathered from a public institution (the public sector drinking water production and distribution enterprise, the national census of 2001), and a number of private institutions (including water cooperatives). The information makes it possible, among other things, to establish, on a geographical basis, the population served by the big system and to match those data with factors such as age and housing conditions. These data were complemented by the results of a questionnaire-based survey about water cooperatives carried out by the World Bank, which were processed and integrated into the database.

In addition to the database, research included a questionnaire and a series of interviews and observations carried out among a sample of users of the cooperatives, in order to study more precisely the organization of small systems. The questionnaire was elaborated in order to reflect as accurately as possible types of accommodation inhabited by the subjects (building materials, number of rooms), their social characteristics (academic qualifications, main job) and their social and political practices (involvement in the life of the neighbourhood or the cooperative), and to compare them with their uses of water and views about the service provided by cooperatives (opinions about water quality, rates defined by cooperatives, etc.). A twin two-stage approach was implemented, involving a selection of the neighbourhoods most representative of the diversity of the agglomeration's inhabitants, and the replies to the questionnaire of 10 per cent of the households referenced (550 questionnaires were collected from the agglomeration of La Paz and El Alto, 50 of which from the cooperatives).

The socio-technological system of La Paz–El Alto

The concession of La Paz and El Alto

If El Alto became administratively autonomous from La Paz in 1985, both cities have been equipped since the 1960s with a range of services covering the whole agglomeration. In fact, the potable water production and distribution service is one of the oldest urban services provided to inhabitants: in the early 2000s, 81 per cent of households in El Alto and 85.7 per cent of households in La Paz were connected to the water supply. In spite of sporadic demands that have the effect of modifying its administrative architecture, the big system has been operated since 2008 by a single public enterprise set up as a limited company: EPSAS.SA (Empresa Pública Social de Agua y Saneamiento, Sociedad Anónima). This enterprise organizes the production and distribution of drinking water without, technically speaking, making a distinction between the two municipalities (Hardy, 2009). The organization is promoted by decision-makers as the very model of a modern water and sanitation service. However, in spite of the fact that the network's coverage has increased, it still does not serve the entire population. The neighbourhoods in the agglomeration's most recently built outskirts have not benefited from the extension of the network, and some neighbourhoods are forced to use relatively unsophisticated technologies (Poupeau, 2010). The most recently built neighbourhoods have a lower connection rate than their more established counterparts. These small systems generally capture fairly weak flows of water from springs or wells which can be rendered drinkable and distributed locally.

The coexistence of two forms of drinking water production and distribution thus raises the question of what kind of socio-technical model residents want to see implemented. This differentiation provokes tensions that occasionally degenerate into social conflicts, as in 2007 when social organizations of El Alto reclaimed the withdrawal of the private operator in order to restore a public service that would provide "water for all". There are also conflicts between municipal administrations, most of them over questions of territorial boundaries, as well as conflicts between residents of border areas torn between the prospect of joining the municipality of La Paz in order to benefit from urban services or staying in more rural communes, less well equipped but less costly in terms of local and property taxes. These social tensions are exacerbated by the imbalance between the two municipalities: although political and economic decision-makers organize urban services on behalf of the entire agglomeration, all forms of capital – economic, cultural and social – are concentrated in La Paz, which is able to use them for its territorial development and geographical extension plans.

The coexistence of different drinking water production and distribution models raises the question of just how complementary those models are. It is necessary to take into account the constraints of the national Bolivian context upon the agglomeration of La Paz–El Alto, constraints that explain urban services policy. For much of the twentieth century, La Paz was the country's largest city, but its hegemony was gradually undermined, on the one hand – at the national level – by the eastern city of Santa Cruz de la Sierra, which since 1950 had attracted an increasing number of economic migrants from the Andes plateaux; and on the other – in terms of the agglomeration – by El Alto, transformed into a receptacle for urban growth by the scarcity of building land in its immediate neighbour.[2] Faced with an increasingly competitive economic situation, authorities from La Paz decided to focus on modernizing water services, part of it with the help of German aid money.[3] The model of water service covering a very large percentage of the population has been used by municipal authorities in order to promote an image of modernity, and to limit the city's loss of attractiveness at the price of increasing the cost of water.

The big system at the heart of urban development

The reasons why production and distribution of drinking water in the La Paz–El Alto agglomeration is managed by a single company can be explained in reference to the history of the big system currently in operation: in 1906, a network of wells and springs scattered around the urban area were developed with a view to reproducing the European hygiene-based model of the production and distribution of potable water. One of the main problems facing water managers at the time was the pollution of rivers by the mines in the mountains overlooking the city (Municipal Archives of La Paz, Municipal Bulletin, 1902–1908). In order to obtain the objective of providing the Bolivian capital with a drinking water provision system regarded, at the beginning of the twentieth century, as modern, the authorities did not hesitate to recruit a European engineer from Germany, tasked with developing the drinking water provision model. The municipality of La Paz managed five separate gravity-based systems scattered around the urban area. With the founding of the municipal enterprise, SAMAPA,[4] in 1966, on the advice of the German government's overseas aid department, which also provided the finance for the first water treatment plants, three technological systems were introduced, each guaranteeing the chain from the raw resource to the distribution of drinking water – the subsystems Hampaturi-Pampahasi, Milluni-Achachicala and Tilata-El Alto. These subsystems gave birth to the contemporary big system (see Box 8.1).

Box 8.1 The big system: La Paz–El Alto

The operating company, ESPAS, captures water from rain runoff and, more rarely, glacier runoff, for the municipalities of La Paz and El Alto and the neighbouring rural commune of Pucarani. Nevertheless, capture, treatment, storage and distribution operations explain why the big system in La Paz–Alto is still divided into three subsystems. Water captured at Hampaturi is transported to the Pampahasi water treatment plant in La Paz before being entirely distributed via the network to the southwestern sector of the municipality (the Hampaturi subsystem supplying approximately 272,000 residents). On the other hand, water captured in Milluni is treated at the Achachicala plant and almost entirely used to supply central La Paz. However, some water is deviated upstream to be treated and distributed by the El Alto system (the Achachicala subsystem supplies some 284,000 residents). Finally, water captured in Pucarani is treated in El Alto where it is fed into the city's potable water distribution network, which also supplies the neighbourhoods located on the western slopes of the municipality of La Paz. In El Alto there are also 30 wells that capture water from the groundwater table which is then treated in the Tilata plant before being fed into the El Alto network (the El Alto subsystem, supplying 915,000 residents).

The implementation of the big system model in the early decades of the twentieth century made it possible to produce and distribute very large quantities of high-quality drinking water. While, in 1966, only 10 per cent of the agglomeration's 495,000 inhabitants had access to drinking water delivered via the municipal network, in 1982, in spite of an increase in population, 65 per cent of residents were connected. In the 2001 census, approximately 84 per cent of a population of 1.5 million was connected to the big system, which for a city in the "South" is an impressive figure in terms of a project promoting "modernity" through the development of an urban service. This technological system also has the advantage of providing an instrument of territorial control.

The implementation of small systems

With the development of the big system, the small systems that once served the various peri-urban communities were gradually absorbed by the expansion of the agglomeration. However, they did not disappear. At first, they did not seem to constitute an issue for the agglomeration, and both the water company and the municipal authorities took no interest in them. For example, only 3.6 per cent of the 165,320 households questioned in the 2001 national census declared that they accessed water using this method. The figure may seem low in comparison with the 84 per cent of households connected to the big system, and, in consequence, to the remaining 16 per cent, but this discrepancy can be explained by the vagueness of the answers to questions posed in the census, a factor associated with the social characteristics of those using the small systems and their geographical location in the city. In reality, around 80 small systems were still operational in 2008. They remain very difficult to identify: by law, they have to be registered by their operators with a department of the Ministry for Water, but few of the complicated and costly bureaucratic steps are required.

First, small systems are to be found in "non-constructible" urban zones, notably the *laderas*, or western slopes of La Paz, inhabited by people from the country who migrated to the city during the period 1960 to 1980. The explanation for this phenomenon is, primarily, to be sought in reasons of a technical nature. The water company is not legally authorized to extend its network into areas threatened by natural catastrophes such as landslides. Indeed, nobody is supposed to live in such areas, but since urbanization is not strictly monitored, entire neighbourhoods have been developed and their inhabitants who, having migrated from the towns and villages of the altiplano, did not have the financial resources to move to anywhere but these expanding informal neighbourhoods, and found in the small systems a solution to their water provision needs.

Unsurprisingly, small systems also operate in recently urbanized zones, namely the peripheries (semi-urban areas of El Alto and La Paz). The urbanized space is full: it can therefore only be extended by taking land from neighbouring municipalities (Urquizo, 2009), through building on increasingly dangerous sections of the *laderas*, or, lastly, by constructing multi-storey buildings (only authorized by the municipality as recently as 2010). Furthermore, El Alto is spilling out beyond the road systems defining its limits, on the one hand, into its own rural areas, thereby regularly obliging the municipal authorities to declare once rural areas urban, and, on the other hand, into neighbouring municipalities where urbanization is still less strictly controlled than it is in La Paz, or even El Alto. Because local authorities are almost entirely unable to control urban expansion, they are unable to provide the necessary infrastructure. Here again, people living in those areas have opted to move to an urban area and put up with the temporary lack of the services normally associated with it: if they cannot be connected to the big system, they turn, in the meantime, to the solution offered by the small system.

Finally, some small systems are located in urban zones served by the big system. Unlike the poorest residents or those who live furthest away from the urban networks, people living in these areas have chosen to obtain their water from a small system, due to the considerably lower costs offered. Because of these varying situations, small systems are characterized by an impressive variety of legal statuses ranging from water committees, whose members include neighbourhood worthies, to cooperatives with more formal legal statuses. In fact, small systems cover a relatively large area of the agglomeration and demonstrate the degree to which access to the big system and to smaller systems serves as an indicator of socio-geographical inequalities.

Small production and distribution systems: indicators of precariousness

The big system was developed by gradually integrating small systems for the production and distribution of drinking water. Mirroring this approach, small systems should, logically, follow the same arc and become subsumed into the big system. However, for at least two reasons, it seems that this evolution does not occur. The first is linked to the technological and operational aspects of the big system, a gravity-based system that cannot be extended due to the agglomeration's topography – La Paz is located in a valley 3,600 metres above sea-level at the foot of the Cordillera Royal, while El Alto is situated on a plateau that rises to 4,000 metres in some of its peripheries. In fact, the big system already comprises three subsystems designed to confront the technical difficulties involved in extending the network.

The second reason is linked to the comparative cost of water supplied by the big system and the small systems in the context of urban growth. Indeed, the precariousness of households in the new neighbourhoods and the marginal cost of access to the networks tend to have the effect of increasing the number of small systems. In the big system, the price per cubic meter is 1.9 bolivianos when monthly consumption per household is under 15 cubic metres, and 2.6 bolivianos when it is over 15 cubic metres. The detailed questionnaire-based survey on water consumption carried out in the whole agglomeration of La Paz and El Alto between 2007 and 2009 reveals that in the neighbourhoods on the periphery characterized by precariousness and casual labour, average water bills amounted to a under 5 per cent of the total household budget[5] (or, in other words, a bill of around 20 bolivianos in a household with an average income of between 350 and 400 bolivianos per month). In the central areas of La Paz, a household connected to the big system receives an average bill of a little over 30 bolivianos, which accounts for around 2.5 per cent of a monthly budget of €1,200 (the minimum wage). In comparison, and taking into account the wide diversity of operators of small systems at the national level, the World Bank survey estimates the average price of water at ten bolivianos per month per household, with an average monthly consumption rate of ten cubic metres.

There are more customers of small systems in La Paz (8.6 per cent) than there are in the agglomeration as a whole (3.6 per cent). Although this phenomenon may, at first sight, seem counter-intuitive, it can be explained by the incomplete nature of the data alluded to above. The small systems that have developed in La Paz are older, better known and have a closer relationship with the municipality; indeed, as we will see, they are used in certain areas as community-based strategies to reduce the cost of the service. In total, 23,160 connections are assured by 47 local operators and/or OLPEs (Operadores locales de pequeña escala: "Small-Scale Local Operators"), a figure that contrasts with the 160,000 households served by EPSAS, the operator of the big system covering the agglomeration as a whole. Applying a participatory approach, these committees, cooperatives and non-profit-making associations attempt to provide water to local people at low cost.

In order to add to the data provided by the geo-referenced information system and test the hypothesis that the systems complement one another, the research has focused on a particular neighbourhood (Cotahuma), the characteristics of which make it possible to address the most delicate points of the analysis. Located on one of the slopes of La Paz, Cotahuma's population is not particularly poor, a fact that makes it possible to study reasons for using the local cooperative, Las Nieves, other than meeting the most basic needs. Above all, the monographic approach allows for a detailed analysis of the uses to which water is put and the costs of those uses, an area that is not covered by approaches applied by international institutions.

The social conditions of the self-organization of the service

The neighbourhood of Cotahuma

The neighbourhood of Cotahuma is situated between a rocky, 100-metre wall and the Calle Buenos Aires, one of La Paz's main centres of economic activity. Cotahuma started expanding in the 1970s, with the construction of homes, workshops and businesses built without adequate technical criteria on unstable, sloping terrain. The construction of houses that did not observe town planning rules or respect the environment, along with the gradual densification of the neighbourhood, contributed to increasing levels of risk, making it more difficult to install the big water system. Most of the neighbourhood obtains its water supplies from a highly developed system of around 30 cooperatives that have been operating for decades. Only 35 per cent of the residents of Cotahuma living south of Calle Buenos Aires are supplied by the big system run by the public enterprise. The efforts undertaken by the neighbourhood's cooperatives, with the help of external institutions, to provide high-quality water to their consumers culminated in the 2000s with the setting up of a Water Committee.

The involvement of members in operating the cooperatives in Cotahuma and the relative proximity of the leaders of these organizations helps prevent issues concerning the service from degenerating into disputes. Problems are solved in meetings at which all parties have an equal say, unlike private business meetings where the influence wielded by shareholders depends on the number of shares held. Furthermore, working with the Pan American Health Organization, present in Cotahuma since November 2008, has increased the awareness of leaders of cooperatives of the need to monitor the quality as well as the quantity of water distributed. At the time the survey was carried out (2008–2009), 16 per cent of the agglomeration's cooperatives were equipped with a system enabling them to provide residents with water adjudged by the municipal laboratory to be of acceptable quality. Approaches applied by cooperatives to improving water quality include thoroughly cleaning reservoirs, installing water chlorination systems and purchasing test systems designed to measure water contamination (chlorine and pH levels).

From a technical point of view, the support provided by the Pan American Health Organization has played a crucial role in terms of explaining methods for improving water quality. But the organization also delivered social support in a series of meetings with managers of the neighbourhood's cooperatives who did not know each other well, if at all. These meetings led to the setting up of a Water Committee encompassing all these entities. "In Cotahuma, I met a community", said the engineer responsible for the PAHO programme in the second water seminar, *Agua y Salud* held in March 2009. The leaders of the cooperative wanted the relationship with the Organization to continue: "That's what we were missing", said one of them during a visit to the installations, thanking the engineer for guiding them through training days and visits to cooperatives to evaluate piping and chlorination systems.

La Paz city municipal government was frequently accused by cooperative leaders of providing insufficient support. Indeed, some went so far as to claim that "the municipal system is the enemy of the cooperative system". They pointed to a lack of material, financial and moral backup. In effect, as the PAHO engineer recalls, "the municipal government does not recognize the cooperatives". This makes it impossible for the city authorities to provide any aid. However, thanks to the personal efforts of the director of relations with social organizations at City Hall, some support is forthcoming. When the Water Committee was being set up, he insisted on the necessity of implementing an action plan with precisely defined objectives, producing an assessment of work undertaken, and organizing further training days with a view to sharing technical knowledge.

Furthermore, the President of the Water Committee highlights that "cooperation leads to integration", insisting on the necessary involvement of consumers in the life of the cooperative. Cooperatives encourage the participation of the residential community, a collective approach that mirrors the forms of cooperation and self-organization promoted by Elinor Ostrom in her critique of the "tragedy of the commons". What, then, are the social foundations of a collective approach to management involving residents, local water resources and the services associated with them? After studying the cooperatives operating in Cotahuma through the prism of interviews with their leaders and with representatives of external institutions (PAHO, municipal laboratory), and of public meetings (seminar on water quality, meeting to set up the Water Committee), the research focused on a cooperative that applied a particularly collective approach and highlighted the social conditions in which it operates.

Individual and collective water consumption: "water brings us together"

Founded in 1972, the Las Nieves Ltda cooperative, an organization recognized by the National Institute of Cooperatives, provides water services to the 186 households in Cotahuma's Alto Tembladerani neighbourhood.[6] The cooperative is run by seven permanent members, and elections for leadership posts, in which all members must take part, are held every two years. According to the current President of Las Nievas, the neighbourhood is characterized by pronounced social homogeneity in that most residents are self-employed tradespeople (mechanics, plumbers, shoemakers, clothes makers, etc.). This claim was validated by the questionnaire (see Box 8.2). However, he does state that around 15 families are from a higher social strata (lawyers, engineers, professors, etc.), which would appear to confirm the area's social diversity. Separated by no more than a few yards, brick houses several storeys high, freshly painted and decorated and adorned with pot plants, stand alongside dilapidated cob dwellings. Apart from the residences of the better-off, the neighbourhood's population resembles that of the expanding peripheral neighbourhoods of the city.

Members of the cooperative have specific rights and obligations regarding the management of the cooperative. In exchange for attending meetings, voting in elections for directors and permanent members, and helping maintain and repair equipment, members have the right to make decisions, and all the cooperative's members are equally important. The fact that almost all the

Box 8.2 A modest neighbourhood

A questionnaire was sent out in March/April 2009 in the neighbourhood of Alto Tembladerani which was used to complement available data from the 2001 census and the geo-referenced database. The residents of the neighbourhood present a modest profile similar to that of other neighbourhoods on the *laderas*, the emergence of which is linked to the rural migration of the 1960s and 1970s. This factor explains the predominant material out of which homes are constructed: 72 per cent of homes are made of adobe, a cheap material used in rural areas. Twenty-two per cent of the families surveyed live in a single room, while 45 per cent inhabit more than three rooms. Half of the households in the neighbourhood comprise four or five people, but 30 per cent include six people or more. Relatively few people live on their own. From a socio-demographic point of view, 61 per cent of heads of families are homeowners and have three children or more (78 per cent of children are under 15), while 28 per cent of heads of families still live with their parents.

The socio-professional characteristics of the heads of families confirm the modest situation of the neighbourhood: 22 per cent are tradespeople while 28 per cent are office workers or employees in the services sector, where they earn low wages; 17 per cent are unemployed and 11 per cent are illiterate; if 51 per cent of those surveyed left the education system at secondary school, only 16 per cent went to university. Most people came to live in the neighbourhood for family reasons (44 per cent), but many are also encouraged to live near the city for professional reasons (43 per cent). Sixty-two per cent of those surveyed were born in La Paz or its province, but only 22 per cent were born in the neighbourhood itself. All of those surveyed had been living in the neighbourhood for at least five years. While 72 per cent of the families who answered the questionnaire speak Aymara, only 22 per cent think of themselves as belonging to an indigenous people.

If 100 per cent of residents are connected to the cooperative's water network, only 78 per cent have a water supply inside the home and 55 per cent have personal toilets connected to the drainage system (45 per cent of residents use communal restrooms). From the point of view of bodily hygiene, 33 per cent of families stated that they wash their children more than twice a week and 56 per cent of adults claimed that they wash twice a week, as they have to use public showers. All those surveyed use water delivered directly to their homes to do the washing up and eject wastewater through the plumbing system. None of them use rainwater for cooking. However, problems persist: 17 per cent say that their children have had gastric problems.

consumers are also members suggests that the cooperative is informed by a desire to be make them responsible, involve them, and foster a collective *esprit de corps*. As the director of the cooperative observed: "Water brings us together".

This involvement in the life of the cooperative goes hand-in-hand with the high level of satisfaction expressed by consumers. There were, for example, no serious complaints about the monthly water bill: 95 per cent of those surveyed think that the price for water (five bolivianos per month, or, in other words, less than a euro) is reasonable, and 72 per cent think that the price of connecting to the service offered by the cooperative is affordable (500 bolivianos, or slightly less than €50). In comparison to the average monthly income of a household of employees – €1,200 – with an average monthly cost of 2.5 per cent of that budget (approximately 30 bolivianos), the monthly price of water supplied by the cooperative is less than 0.5 per cent of the monthly income. But it is in comparison to the expanding peripheral zones not served by the big system, where the monthly price of water accounts for 5 per cent of the total monthly household budget (see above) that, from a financial point of view, the cooperative system really comes into its own, claiming a mere 2 per cent of that budget. Finally, the cooperative enables households with low or irregular incomes to pay similar or lower rates for water as other households. It would appear that the lower the income of households, the more attractive the prices offered by the cooperative.

Although 78 per cent of those who answered the questionnaire were aware of the fact that the province of La Paz is encountering problems with droughts and that water is becoming increasingly hard to come by, 83 per cent claimed not to have observed any signs of scarcity in the Cotahuma neighbourhood. The perception of the service provided by the cooperative is relatively positive, and, while 83 per cent of the inhabitants have suffered water outages, the majority of respondents (67 per cent) recognize that they occur very infrequently (cleaning the reservoir, etc.). Furthermore, for 78 per cent of the sample, the water provided by the cooperative is of a good quality, while for the remaining 22 per cent its quality is average. Thus, users express overall satisfaction with the service, even though the director of the cooperative recognizes that the water system needs to be optimized in order to meet demand from residents of nearby neighbourhoods currently connected to the public service. Similarly, he is thinking of installing water counters in all households in order to regulate levels of consumption, and of introducing contracts between users and the cooperative in order to limit abuses by landlords who make their tenants pay over the odds for water services.

Advantages and drawbacks of the community-based approach

In the survey, directors highlighted a decrease in participation on the part of members of the local community at Las Nieves and in the other cooperatives studied. The "younger generations", which are becoming increasingly visible in the neighbourhood (41 per cent of the heads of families are under the age of 40), stand accused of abandoning traditional practices and even of being less willing to learn the Aymara and Quecha languages spoken by the so-called "indigenous" inhabitants of the altiplano who accounted for most of the rural migration to the La Paz *laderas* in the 1970s and 1980s. A growing number of cooperatives are considering increasing the price of water for consumers who do not help with maintenance or participate in general meetings. Abuses committed by certain residents are also mentioned: more and more presidents of cooperatives point to the need to install water counters to measure consumption, not necessarily with a view to introducing prices per cubic metre, but at least to charging consumers who overstep the mark. However, since a single water counter would cost at least $50 (350 bolivianos), cooperatives would be forced to increase their prices in order to install them.

From the point of view of younger consumers (between 25 and 30 years old), helping out with the work of the cooperatives is seen as a chore and, in terms of their lifestyles or, at least, the lifestyles to which they aspire, as a meaningless activity. Material access to domestic water, the quality of the service provided and a desire to do things outside the neighbourhood are also mentioned in this regard. For young people with jobs, the deterioration in working conditions, whether for employees, small entrepreneurs or workers in the informal sector (street vending, building sites, etc.), makes it hard to make a commitment to collective action. For example, at Las Nieves, members not only have to attend meetings but also work on equipment (repairs, cleaning, etc.) at least 15 Sundays a year in order that individual families contribute approximately 500 bolivianos ($80) worth of labour annually. Family contributions are recorded in a notebook by the president of the cooperative. The president has the right to demand sanctions at general assemblies. However, in practice, only price rises are ever envisaged, a demonstration of the increasingly generalized disaffection with the collective.

The relative apathy of younger residents highlights another precondition of the existence of cooperatives. These small operators have succeeded in providing the neighbourhoods under consideration with potable water and sanitation at low cost, but a lack of funding impacts upon their efficiency: infrastructure maintenance, increasingly demanding water quality standards imposed by the cooperative federation, and the need for all operators to introduce a water treatment system all imply the need to find external support – from city government or NGOs – in order to fund water "professionals" able to supply technical services.

A lack of recognition of small companies in the context of a fragile big system

It is a commonly held belief in the La Paz and El Alto municipal administrations that the production and distribution of drinking water is a complete technical cycle and that this cycle should be managed by a single operator using the approach established by the European hygienist engineers of the nineteenth century. This big system model presents a number of advantages. It supplies a large volume of water, monitors its quality and manages its distribution among a large number of residents. It has proved its efficiency by rationalizing its technical approach, notably thanks to the expertise of the French operator which ran it between 1997 and 2007.

However, the big system is becoming increasingly inefficient in the two cities. In 2008, the national and municipal governments, which had taken the service over from Suez, demanded that the operator of the big system provide services to the entire agglomeration (except for areas at risk of natural catastrophes) within a defined budget. This policy obliges the operator to extend its distribution network in a context in which planning regulations are too frequently ignored. Therefore, the cost of extending the network is only just covered by the financial equalization scheme introduced by the operator to encourage transfers between consumers in wealthier areas and those in poorer neighbourhoods, and by public subsidies. Extensions are often carried out to the detriment of the maintenance of existing infrastructure, provoking a general deterioration of the service and, consequently, discontent among consumers, especially those paying the highest rates. Finally, when consumer contracts are not altered by the change in legal status of the remunicipalized company, approaches to infrastructure similar to those of private operators may be observed: expanding peripheral neighbourhoods are equipped with less infrastructure, of a lower quality, with often failing compensation solutions (condominial systems, involvement of residents in the construction of infrastructure, etc.).

Furthermore, the policy focusing on extending the network implies an increase in demand for water at a time when the resource is becoming increasingly scarce and the climate of the Bolivian altiplano is unquestionably changing. Due to global warming, glaciers are increasingly falling short in their role as retainers and restorers of water, which explains the decrease in the amount of water available during the winter dry season (Carey, 2010). In order to confront this situation, managers follow the logic of the big system. They guarantee supplies by sourcing water further and further away from the area in which it is consumed and stocking it to ensure that the resource is available in periods of low rainfall. This management approach requires investment in costly infrastructure (dams, pipes for transporting water, etc.). The cost of drinking water produced in this way increases, obliging users to reduce consumption, which puts the investment and operating capacity of the company running the service at risk.

By taking such an approach, the operator of the big system effectively ignores the issue of the uses to which water is put by consumers. Originally, the big system was designed – and this is its strong point – to supply enormous quantities of drinking water in a historical context in which the poor quality of available water had a significant impact upon the mortality rate of the population. Today, water is used for different purposes in La Paz and El Alto. Most potable water produced by the agglomeration's big system is used not as drinking water but, instead, is put to industrial and commercial as well as domestic use (cleaning homes, watering gardens, etc.). Such uses rarely demand that the water they require be drinkable. But the big system, precisely because it is a unified model, is not capable of distributing drinking water and water suitable for other purposes through different networks. Its inability to support the cost of such differentiation impacts negatively upon its efficiency.

In a context in which the big system for the provision of drinking water is declining in terms of efficiency (scarcity of the resource, operating costs, etc.), small systems offer a genuine alternative. They already enable a substantial percentage of the population of the agglomeration to procure water at very low prices, without undermining the big system, the current capacity of which has little room for expansion. However, the small systems encounter obstacles in terms of implementation. The profile of their consumers, which is similar to that of consumers requesting to be connected to the big system, demonstrates that the cooperative approach is not underpinned by political principles, but by a form of instrumental rationality designed to provide access to water in order to ensure that people live as comfortably as possible in a given urban milieu.

Without question, management approaches applied in the small systems favour the emergence of a certain degree of autonomy on the part of their members, who are encouraged to manage the resource themselves. On the other hand, this form of self-organization also makes them responsible for the sustainability of the system (Ostrom, 2009). Participation does not guarantee the longevity of small systems: at a time when cooperatives are "professionalizing", younger people tend not to want to help out on a day-to-day basis (maintenance, meetings, etc.), as the Cotahuma case illustrates. This is demonstrated in the monitoring and training of local committees and cooperative managers, in which costs associated with the application of new health standards complicate the management of small systems seeking recognition from local authorities.

However, the advantages for urban services managers of building a bridge between the two water provision systems is illustrated in a number of different areas. In the case of a catastrophe damaging the big system, the operator would not have to provide water to users of small systems. Indeed, the small systems apply technically simple approaches and the problems with which they are confronted are often easy to resolve. Water outages are rare. Furthermore, small systems that deliver sufficient quantities of water can even supply, on a temporary basis, local facilities ordinarily served by the big system, such as schools and health centres, thus avoiding a crisis that could easily spread to the rest of the agglomeration. Small systems are therefore crisis management resources, alternatives to the big system. The coexistence of two systems facilitates urban delivery, not only by rendering it less fragile but also by providing a potential source of territorial solidarity in terms of the management of crisis situations. It is even legitimate here to talk about the possibility of "an evolution toward composite systems hybridizing conventional big systems and alternative systems designed to function at smaller geographical scales" (Coutard, 2010: 104).

Conclusion

Cooperation between big system and small system managers creates opportunities in terms of developing new approaches to effective management: the complementary nature of the relationship between the big system and small systems provides a testing ground for the development of alternative solutions in situations in which urban growth is accompanied by an increasing scarcity of natural resources, rises in prices of basic services and the need to consume differently. However, cooperatives appear as a complementary service to big systems when they are likely to foster the emergence of self-organization of water service in their area. If community engagement ceased to exist, it is feared that a shift into more informal practices would occur and, if so, that would not provide the same guarantees in terms of water quality or water supplies in a situation of crisis, as the cases of La Paz or El Alto have already showed. One may wonder also if these informal practices would be sufficiently effective in tackling the looming effects of climate change on water supply to Bolivian cities.

Notes

1 In February 2008, the French Institut de Recherche pour le Développement (IRD) signed a scientific and technical cooperation agreement with the Autonomous Municipal Government of La Paz (AMGLP) in order to study urban risks in the city of La Paz in Bolivia (Hardy, 2013). Furthermore, between 2006 and 2010, the Institut Français d'Etudes Andines (IFEA, CNRS-MAEE) conducted research on inequalities in access to water in Bolivia. Researchers and students, Bolivian and French, working for these two institutions, collected various data on inequalities in access to water in the city of La Paz and El Alto. A participatory approach was applied to identifying information to expand the database. The team, made up of researchers active in the social sciences, as well as engineers and technicians specializing in GIP, organized theme-based meetings involving specialists from the public and private sectors in each area, and also representatives of institutions involved in the management of the water service. These meetings made it possible to draw up a list of water resources and to localize available information. For more detail on the maps see Hardy and Poupeau (2014: 89, 93 and 94).

2 The effects of the 1953 Agrarian Reform, which caused the landownership system to fragment, made themselves felt in the migrations of the following generation commencing in the 1980s (Albo *et al.*, 1985).

3 Since the late nineteenth century aid money from Germany has played an important role in public sector development projects in Bolivia; historically, this aid has been linked directly to the presence of numerous German companies in the country, notably in the mining industry (Eichholz, 2010).

4 SAMAPA (Servicio autónomo municipal de agua potable y alcantarrillado) is a decentralized municipal enterprise responsible for urban water services and sanitation set up in 1967 by Municipal Amendment No. 7597. It has a legal identity and its own patrimony, and enjoys operational and administrative autonomy. After a transitional period of four years, the municipality of La Paz definitively transferred all its water services to SAMAPA.

5 Compared to the monthly cost of basic services, the bill accounts for 7 per cent of the household budget. This population uses monetary income to pay for basic services,

while procuring other services on a non-cash basis. This means that the ratio between the water bill and the total budget cannot be interpreted as it is in industrial companies (Poupeau, 2011b).

6 Since consumers are generally also heads of households, each of which has an average of five members, it is estimated that the cooperative supplies approximately 1,000 people.

9 The decentralization of supply in four Uttar Pradesh municipalities

Rémi de Bercegol and Shankare Gowda

In India, where, at the beginning of the 1990s 37.7 per cent of households were without access to water, the water distribution service has long epitomized the failures of a centralized management system. The public service management framework has since been redefined following the constitutional recognition, in 1992, of local governments (through the seventy-third and seventy-fourth amendments for rural and urban areas, respectively), to whom a number of responsibilities were devolved, including management of the water distribution service. However, as, according to the 2011 Census, 53.4 per cent of India's population still has no water at home, it consequently seems reasonable to question the effectiveness of this reform. This chapter outlines the challenges associated with decentralization[1] of the water service in India's small towns, which commonly lack urban facilities and infrastructure (Kundu and Thakur, 2006): How has water service management changed in this type of town? Has decentralization helped increase access to water? What are the barriers to universal access to the service in small towns in India?

After reviewing the main features of the decentralization reforms, small towns and water service accessibility in India, this chapter will present the findings of studies (Bercegol, 2012) undertaken in four small municipalities in Uttar Pradesh, one of the most densely populated and poorest states in India. First, we will look at the main empirical reasons behind the municipalities' technical inability to take over management of the water service. Then, from analysis of the different modes of access in each of these towns, we will examine the limited scope for action these municipalities actually have for improving water service performance. Finally, we will demonstrate how the substantially different levels of infrastructure available in these small towns, which nevertheless all have similar needs, is evidence of the ongoing major role being played by the provincial government, despite decentralization. By providing an up-to-date insight into a largely unreported urban situation, that of small municipalities (Satterthwaite, 2006), the aim of this chapter is to contribute to the current debate on changes to the water service in a highly diverse urban world (Dupont, 2002).

The water service in Indian small towns

The decentralization reforms in India

Within some of the specialist literature on drinking water issues, it is agreed that the failure to provide universal access to the water service in India stems not so much from restrictive demographic pressures (1.21 billion inhabitants in 2011) or unfavourable geo-climatic conditions (veering between severe droughts and heavy monsoon rains), as from the fact that institutions are highly ill-adapted to deal with the complexity of local situations (Saleth, 1996). Through the major reform launched at the beginning of the 1990s as part of the seventy-third and seventy-fourth amendments to the Constitution, Indian legislators hoped to avoid the pitfalls previously encountered during implementation of an essentially centralized management model by granting local governments a positive role to play.

Unlike certain other countries that came under international pressure to change, the decision to introduce decentralization in India was instead made as part of internal/national policy and was the outcome of a long-standing debate on the role of local institutions (Tawa Lama-Rewal, 2007). Indeed, this reform resurrects the myth of the Panchayat (which literally means "assembly" – *ayat* – of "five" – *panch*), the oldest type of local governance in India that took the form of a small village council made up of respected elders and which remains deeply ingrained in the collective psyche. Devolution to the municipalities of not only political power (with the introduction of municipal elections), but also technical responsibilities (for example, the management of certain basic public services, such as water distribution), has thus given rise to great expectations of change. However, nearly 20 years after its introduction, initial evaluations of this reorganization of authority in India remain patchy. Thus far, analyses have mainly focused on political representation within local authorities (Mathew, 2000), but have paid less attention to the technical capacity of these authorities to assume the new roles allocated to them. In addition, a number of outstanding questions remain concerning the ramifications of these reforms for different-sized towns, particularly the smallest.

Small towns in India: an overlooked issue

When dealing specifically with devolved public services, analyses most often focus on the major metropolitan areas of Mumbai, Delhi, Calcutta or Chennai (Ruet *et al.*, 2002), thus indirectly helping to substantiate the impression of urban India as being, above all, a very large city, that of the *Maximum City* (Mehta, 2004). However, the complexity of an urban system that consisted of 7,935 towns and cities in 2011 cannot be defined by analysis of the few cities in India with over a million inhabitants alone. A brief review of the country's urban hierarchy shows that the distribution of the urban population (only 27.92 per cent in 2001 and 31.16 per cent in 2011[2]) is far more complex than the simple urban/rural dichotomy would suggest. In 2011, 7,438 of India's 7,935 towns had fewer than 100,000 inhabitants and preliminary results from the census indicate that many of these are small towns, including 2,774 new towns[3] that were still considered rural during the last Census in 2001 (Bhagat, 2011). In the shadow of the "India shining"[4] metropolises, there is another urban India: that of smaller towns, the statistical accounting of which has been the subject of much debate (Kundu, 2011) due to their being in transition. Although their existence can no longer be denied, these small towns are no longer "rural" but neither are they yet fully "urban" (Denis and Marius-Gnanou, 2011).

It remains difficult to identify precisely this type of agglomeration, as there is no real agreement on the definition of the term "small town", which can have anything between 20,000 and 500,000 inhabitants (Dhaliwal, 2004). Compared to cities with over a million inhabitants, the vagueness surrounding this category has a detrimental effect: "small towns" have far fewer resources than large cities; poverty is more widespread, particularly in towns with fewer than 50,000 inhabitants; and the percentage of the population without basic services is much higher (Himanshu, 2006; Shastri, 2011). Moreover, these variations are even more pronounced in the least developed regions, especially those that have witnessed rapid urban development, such as Uttar Pradesh (Kundu, 2011).

Access to water in small towns in Uttar Pradesh

As things stand, there is a lack of recent and accurate information available in the current literature on the specific situation of small towns in India (Bercegol, 2012). For instance, according to data from Jal Nigam (JN, the governmental agency in charge of water) in Uttar Pradesh, the entire urban population of the region's 623 agglomerations, both large and small, has had access to water since 2004. Similarly, the Indian Census of 2011 confirmed a highly encouraging access rate for these towns of 95.8 per cent. Unfortunately, however, this official version needs to be put into perspective as, although highly positive, it reveals little about the wide range of different situations encountered.

First, these figures aggregate all modes of access, whether through a water distribution system or stand-alone hand pumps and boreholes (not connected to the public network). In reality, only just over half of the urban population has access to running water from the distribution network, with the remaining households obtaining their water from other sources (in 2011, 46.3 per cent used hand pumps and boreholes, 0.8 per cent used traditional wells and 1.3 per cent used surface water). Some 21.2 per cent of households have no access to water at home and, for 5.2 per cent of these households, the nearest water point is more than 100 metres away from their home.[5] After having increased during the 1980s, from 50.76 per cent in 1981 to 58.8 per cent in 1991, the number of households with access to the piped network dropped by 2.32 per cent during the 1990s to stand at 56.5 per cent in 2001. This figure then fell by a further 5 per cent to reach 51.5 per cent in 2011, placing Uttar Pradesh among the states with the lowest connection rates to the water network in India and revealing the heavy dependence on boreholes and hand pumps.

In addition, by grouping towns together in the same simplistic "urban" category, many of the official statistics fail to show differentiated access based on agglomeration size. However, access to the water distribution network remains low, since it is also being hampered by the extremely poor performances of a service for which the distribution standards, set by Jal Nigam, not only vary considerably with the size of the town (between 135 and 150 litres per capita per day in large towns of over 500,000 inhabitants, and between 70 and 135 litres for all other towns since 2002), but they are also rarely met by one-third of agglomerations, particularly small towns (Mishra, 2002). Thus, in 2001, 85 per cent of towns with between 20,000 and 100,000 inhabitants had a service that met fewer than 50 per cent of the standards. In contrast, two-thirds of agglomerations with fewer than 20,000 inhabitants (67 per cent of towns) provided a service that met over 75 per cent of the set standards, as these were easier to attain (70 litres per day). This situation illustrates the problems encountered when progressing from a system supplying a small number of inhabitants, essentially a large village with a water network, to a proper urban water scheme that needs to supply a population of over 20,000.

Finally, since they are based on the quantity of water distributed rather than the quantity actually received, these government figures tend to overestimate the actual situation experienced by the population, as physical losses of between 25 and 40 per cent (McKenzie and Ray, 2009) further impact upon the already poor performances seen above. This data also overlooks other issues, such as intermittent service, which directly affects the water quality: when the service is only available for a few hours each day, the resulting changes in pressure damage the pipes, and the water in the network is polluted by wastewater seeping into the system.

The aim of this study is to examine the specific cases of four small towns in India of similar size, each with a population of around 20,000 inhabitants, and situated in the same institutional and geographical region, namely the eastern part of Uttar Pradesh. In contrast to the west of the state, which in part benefits from its proximity to neighbouring Delhi, the federal capital of India, economic development in eastern Uttar Pradesh is low and poverty is much higher (World Bank, 2002), meaning that there is a pressing need for any reforms undertaken to bring improvements.

Although demographically very similar, each of the small towns studied has a specific feature (cultural, institutional, economic and social), enabling potential differences in implementation of the reform to devolve the water service to be identified. Thus, Kushinagar is a small tourist town with a Buddhist religious site that is currently undergoing development;[6] Chandauli is a district administrative centre whose population doubled during the 1990s;

Siddarth Nagar is the administrative centre of one of the poorest districts in Uttar Pradesh and has been defined a priority area by the government; and Phulpur has no specific feature besides its proximity to an industrial fertilizer plant and its majority Muslim population (the populations of the other towns being mainly Hindu). Based on the studies conducted,[7] the low number of households connected to the water network in all four towns, a figure which did not exceed 50 per cent in 2009 (36 per cent in Kushinagar, 31 per cent in Chandauli, 18 per cent in Siddarth Nagar and 45 per cent in Phulpur), raises doubts over the capacity of the reform to provide universal access to the water distribution service.

Decentralized management that is difficult to implement

Management responsibilities, from the theory to its challenging implementation

Prior to reform of the service, all water infrastructure design and management tasks were the responsibility of Jal Nigam, the governmental technical agency in charge of water and sanitation. Thus, Jal Nigam engineers were in charge of the entire drinking water service, from its technical design through to its implementation and day-to-day management.

With the introduction of decentralisation,[8] however, Jal Nigam had to cede part of its responsibilities, as urban governments[9] were officially put in charge of infrastructure *management*. This included both financial responsibility (contributing to initial investment in service infrastructure, billing users and collecting revenue to carry out technical maintenance) and technical responsibility (installing new network connections, routine maintenance and management of service infrastructure). Nevertheless, the Jal Nigam engineers remain in charge of water infrastructure *design* and the technical extension of primary water mains (water towers, pipework, or even treatment plants in some of the larger towns, etc.), which involves working with a range of various institutions (government departments, the government's electricity agency, private contractors, etc.) before transferring the facilities' management to the local authorities. Jal Nigam describes itself as the public water service "nodal agency" (UIDSSMT Chandauli Project), upon which the municipality has to depend if it is to expand access to the service.

Under the decentralization reform, local political representatives have gained a new status that gives them a key role in service governance. This governance was previously the exclusive preserve of the Jal Nigam engineers, who are now officially required to implement the local authorities' requests. In practice, coordination efforts between the municipality and Jal Nigam often lead to tension or even conflict, which hampers development of the service. Over and above the factors of caste and religion,[10] by making the engineers "feel downgraded", this theoretical reallocation of authority to municipal staff appears to be the main cause of engineers' hostility towards elected officials, who they more or less openly discredit. This latent, and sometimes racist, contempt can be exacerbated by distance, as the engineers do not generally live in the small towns studied. For example, the engineer responsible for Chandauli, whose offices are in the neighbouring town of Mughal Sarai, objects to having to go to inspect the facilities in Chandauli on the grounds that "the municipal managers are uneducated".[11] For the Hindu assistant engineer in Siddarth Nagar, rather than being the result of social or technical factors, the low connection rate seen in the market town of Naugarh (an area of Siddarth Nagar where only 18.8 per cent of households are connected to the network) is in part due to the "very poor hygiene" [*sic*][12] of the Muslim population.

In addition, the engineers feel they have been reduced to a form of servitude by the various political pressures and demands placed upon them, making cooperation with the municipality more difficult. While we were conducting an interview at the Jal Nigam offices in Allahabad, an elected official from the pro-Dalit[13] Bahujan Samaj Party (in power in Uttar Pradesh at the time of the study) came in to "demand" the installation of new hand pumps in his ward. Upon the official's departure from his office, the chief engineer launched into a tirade of criticism about the arrogance of this new local government: "what can you do with these small, second-rate politicians who think they can do whatever they want?"[14]

Finally, the municipality remains largely uninvolved in the water infrastructure design process. On the one hand, there is no real consultation structure put in place for the technical and financial preparation of the project and, on the other, elected officials appear so overwhelmed by the complexity of projects that they prefer to abdicate all responsibility to the engineers. Thus, water network design decisions (network configuration and location, financial and technical maintenance plan, etc.) continue to be made without consulting the primary stakeholders, which goes against the principle advocated by decentralization.

Minimum technical transfer

In theory, the transfer of Jal Nigam-designed facilities to the municipalities should be accompanied by at least some skills transfer: "Jal Nigam organises suitable training as well as capacity-building programmes from *time to time* as per the need of the project" (Nigam, 2005). In reality, the engineers are highly reluctant to have to transfer what they consider to be their responsibilities. In Kushinagar, for instance, the Jal Nigam engineers made an unsuccessful attempt to block the transfer of water network management to the municipality:

> it is suggested that ... the system should be continued to be maintained by the UP Jal Nigam itself.... This is necessary because the town area

> committee will not be able to maintain the scheme because they do not have trained engineers with them. The Kushi Nagar Town is going to be an important International Tourist Center of India in the near future with the efforts of government of Uttar Pradesh and Japan. Hence efficient maintenance of the scheme is a must and has an impact on the tourist coming to the towns.
>
> (Jal Nigam, 1990)

Over and above their justification that the town is a showcase for tourism, this clear show of dissent highlights the Jal Nigam engineers' distrust of the municipalities' technical capacities and their reluctance to provide technical training to local municipal employees.

In general, the municipal technical team consists of between two and four locally recruited employees (one of whom will be a plumber). Although some are permanent staff (paid around 6,000 to 8,000 rupees, or about €100, a month), the majority of these employees, such as the plumber, are paid the minimum daily wage (70 rupees, or under €1, a day). It is far easier for the municipality to afford these salaries than to recruit a qualified engineer, who would anyway prefer the career prospects offered by Jal Nigam (where the starting salary is 11,000 rupees a month – just under €200 – rising to 30,000 rupees – around €550 – depending on length of service). With no prior training from Jal Nigam, the municipal plumber learns "on the job" without necessarily taking all the proper precautions, since he is unaware of what these are.

Local employees have never received any training on public health issues. According to Jal Nigam standards, the water tower tanks should be cleaned with chlorine at least twice a year. However, this directive is not followed in any of the towns studied, where water tanks are only cleaned once every 12 to 18 months on average. In some instances, the town hall has simply neglected to budget for chlorine and the networks are checked only rarely. In the tourist district distribution area of Kushinagar, so many people were falling ill that the municipality eventually ordered a clean-out of the offending tank, in which was found the decomposing body of an adolescent who had committed suicide some weeks before. In addition to these virtually non-existent basic checks, there is also a level of unprofessional nonchalance displayed by some of the poorly trained employees. For instance, in a dismissal letter from April 2008, written following complaints from dissatisfied users, the Investigation Officer in Chandauli accuses one of the employees of "playing cards instead of being at his post to open the valves during distribution hours" (extract from the dismissal letter). Finally, in each of the towns studied, employees fail to properly regulate the pressure when opening the valves, which regularly results in water hammer and causes the weakest pipes to burst.

As none of the municipal employees is trained to maintain and fix the Jal Nigam pipework, leaks can remain unrepaired indefinitely, sometimes for months, until the engineers eventually carry out repairs. This can lead to conflict between the municipal authority and Jal Nigam, as in Phulpur, for example, where there were major leakages on the water scheme north of the town. The municipal-elected officials blamed Jal Nigam for this, claiming they had transferred infrastructure to the local authority that was worn and poorly maintained. The engineers, however, blamed these leaks on the incompetence of the municipal plumbers, who, they claimed, had damaged the distribution pipework either when connecting households to the system or simply when opening the valves. When the district's senior public servant finally ordered the engineers to carry out repairs, work was hampered by opposition from an elected official, who claimed that the new pipework would damage the roads in his ward. After several months, the network was finally handed over to the municipality, but the leaks are steadily reappearing and the local authority has had to resign itself to putting in place a makeshift solution, namely installing public taps to reduce the pressure inside the pipes during distribution hours. This is just one of many episodes in which there has been a lack of understanding between the municipal managers and Jal Nigam engineers.

Intermittent electricity supply

This poor relationship is made worse by electricity rationing and inopportune power outages which prevent the pumps from operating, meaning the tank is not properly filled. The resulting changes in pressure damage the pipes and the water distributed becomes polluted by wastewater seeping into the system. The chronic lack of electricity being produced in India not only results in a highly intermittent water service, but also means that not even the highly restrictive power distribution norms in place in Uttar Pradesh are always respected (eight to nine hours in rural areas, as is the case in Phulpur; 14 hours in towns with specific administrative status, such as the *teshil* or district headquarters, four hours of which is at low voltage). The Jal Nigam agency designs water infrastructure based on theoretical power distribution norms that fail to match the daily reality within the small towns. As a result, the hours of distribution have to be rationed. Not only is this restrictive, but these hours are also extremely irregular due to the unscheduled nature of the outages. In Phulpur, the distribution period alternates on a weekly basis, with power supplied during the day one week and during night-time hours the next. In a letter to the regional electricity department dated October 2007, the Mayor of Chandauli also requested the "replacement of the generator supplying the borehole. As this is no longer working, it is not possible for the municipality to use the pumps to supply drinking water". Intermittent power and inadequate equipment are thus major technical issues for the municipality managing the network and further reduce the municipality's already limited scope for action.

The service provided through the network is inadequate

The number of households with an individual connection to the water network is very low compared to Jal Nigam forecasts (upon which the engineers based the financial plan for the system), thus illustrating that attempts to expand the water distribution service have failed (see Table 9.1). If domestic and free connections and public hand pumps have been installed in strategic locations to compensate for the poor service provided by the network, the reform would sustainably improve the water service.

Access is technically fragmented

Individual household connections are installed using technical and financial criteria that fall largely beyond the municipal authority's control. The monthly household connection fee set by the local authority is between 20 and 30 rupees a month (under 50 euro cents). This is consistently lower than the Jal Nigam recommendations for ensuring financial sustainability of the system, yet it is an amount that inhabitants are able to pay (the estimated median salary in these towns stands at between 5,000 and 7,000 rupees a month, or about €100). In reality, it is the fixed cost of the initial connection that remains prohibitive for users. Not only must they pay the upfront fee set by the municipal committee (between 500 and 600 rupees on average), but they must also cover the set cost of the pipework required to connect their houses to the network (which, in general, comes to a minimum of 300 rupees for six metres of pipe). Plumbing costs rise exponentially the further the house is away from the primary network, while, at the same time, the pressure drops and the quantity of water delivered to the tap falls the longer the pipe.

Comparison of the wards with the most connections shows that connection to the water network is not directly determined by caste or social background but by the geographical location (itself socially differentiated) of the

Table 9.1 The number of domestic connections is well below Jal Nigam forecasts

Number of domestic connections	*Phulpur*	*Siddarth Nagar*	*Chandauli*	*Kushinagar*
Jal Nigam forecasts made at the project design stage	3,357 in 2003	3,000 in 2009	Not available	2,400 in 2009
Actual number of connections recorded by the municipality	1,356 in 2009 (45%)	532 in 2009 (18%)	995 in 2009 (31%)	915 in 2009 (36%)

Source: Jal Nigam forecasts and municipality accounting records.

house, with houses in the town centre more likely to be connected, as they are closer to the infrastructure and water towers. Those people living on the outskirts of the town centre experience technical problems (low pressure) and suffer an added financial burden (higher plumbing costs) when connecting to the water supply system. People moving into the now increasingly densely populated outlying areas are thus likely to be discouraged from connecting to a water scheme constructed prior to the neighbourhood's expansion.

All municipalities have opted to install a number of free public taps (which also runs counter to Jal Nigam's financial recommendations of levying a monthly charge of between eight to 15 rupees per household in Phulpur and 24 rupees in Kushinagar).[15] The location of these community points of access to the water system is determined by the local authority based on social factors, the aim being to compensate for the fact that only a small number of the poorest households are connected to the network (thus the cost, covered by the municipality, no longer prevents connection to the system in poor districts). However, there are also more symbolic and economic reasons behind the choice of location, such as wanting to develop the most important and popular parts of the town (in the area around the railway station, in the administrative district or the tourist centre). The municipality's willingness to expand the service remains hampered by technical factors, as certain outlying wards are still unable to connect to the piped water system due to low pressure at the end of the network.

Users are dissatisfied with the service

In each town, satisfaction surveys were carried out with a sample of connected households[16] to assess service performance. One of the most common criticisms is low pressure at the tap. The dissatisfied users nearly all live in wards some distance away from the water tower, at the end of the network. As the pressure drops the further the pipes are located away from the water tower, certain households receive significantly lower quantities of water than those living near the water tanks. In each town, users also complain more or less vehemently about the poor water quality (caused mainly by the lack of tank maintenance and wastewater seeping into the pipes), and this is reflected in the number of cases of waterborne diseases treated by hospital doctors (interviews conducted in August 2008 and 2009). During the field surveys, one of the elected officials in a ward in Chandauli had even advised residents against drinking water from the network and instead to find other sources, particularly during the monsoon season.[17] Finally, as in many other towns in India, people struggle to cope with the intermittent nature of the service, as water is only distributed between once and three times a day and for one to four hours at a time.

This intermittent service is the direct cause of the water service's performance problems, as it results in only limited quantities of poor-quality water being delivered. In India, continuous water supply remains the exception due not only to local politicians' interventions and the limited availability of power, but also to the technical approach taken by the engineers and a deep-rooted "tradition of intermittent service" (Zérah, 2006).

The development of public "off-network" facilities

In reality, to help offset the deficiencies of the poorly aligned water network, municipalities have come to rely heavily on "off-network" facilities, particularly hand pumps (installed by Jal Nigam but financed by the local authorities). The use of these hand pumps is facilitated by the fact that the groundwater tables are shallow in this Indo-Gangetic Plain region, which is fed by the neighbouring Himalayas. In a letter to the district magistrate dated July 2006, the municipality of Chandauli explained that due to

> power supply issues, the municipality is unable to provide a water service and has had to install 103 handpumps to overcome this problem ... furthermore, although the water tower tank has been repaired, pressure levels in certain wards remain too low.

In a further letter, the Mayor of Chandauli directly petitions the district magistrate to "install handpumps in ward no. 2 as soon as possible as the municipality is not able to provide a water distribution service and Jal Nigam has not responded to our requests". Similarly, in a letter dated July 2008, the municipality of Chandauli urges Jal Nigam to start carrying out "repairs on the handpumps as quickly as possible as these are no longer working properly". As they free up the municipalities from the technical constraints of operating an inefficient network, these pumps have become widely popular among local authorities.

Furthermore, by its very nature, decision-making on the placement of these hand pumps, which are completely free to use, is far more decentralized (being the role of local elected officials) than that relating to the piped network, and is undertaken in response to social and technical factors (to supply water to poor wards and/or wards that are poorly served by the network due to their outlying location) and more symbolic considerations (the same as those used to determine the location of water points, installing pumps in places of importance, such as stations, the town hall or in the most prestigious areas, such as the tourist ward in Kushinagar – sector no. 4 on maps of Kushinagar). Thus, in the wards that suffer the most from supply issues, the municipality endeavours to compensate for the poor service provided through the network by installing more hand pumps, a strategy that can also become a political issue. In the run-up to elections (municipal and higher), elected officials may particularly target hotly contested wards, turning the installation of hand pumps in certain wards into a subject of political competition. This supports other analyses of local political "territorialization" issues created by the water service (Bercegol and Desfeux, 2011). The *Kanshi Ram* programme, a special project launched by the pro-Dalit Party in power in Uttar Pradesh between 2008 and 2012 to install facilities and equipment in Dalit districts (such as ward no. 5 in Siddarth Nagar, which has excellent coverage thanks to the standpipes and hand pumps installed through the programme), also fits into this "territorialization" electoral strategy.

The growth of the private market

As in other towns in India, the level of the water service is so poor that customers have been forced into using costly and time-consuming "compensatory strategies" (Zérah, 1999) that provide unequal access and around which a fast-growing local private market has developed. There are no water sellers (water tankers or bottled water sellers as there are in states with deep water tables, such as Tamil Nadu); however, the private market for the installation of individual household pumps, or boreholes with motorized pumps for those able to afford them, has grown considerably over the past few years. As a result, the owner of a specialist water pump shop in Siddarth Nagar claims to sell "over 500 hand pumps a year" (interview conducted in October 2009). Full installation costs 2,800 rupees (just over €50, whereas the median salary in these towns stands at just over €100) and includes the price of the pump, the approximately 12-metre-deep borehole and pipework of around nine metres in length.

Although they are much more expensive, motorized pumps are becoming increasingly popular. An electric pump and 40-metre-deep borehole costs 26,000 rupees (around €520). This already high cost (which the majority of households are unable to afford) does not include the price of the water storage tank, which varies from between 800 to 4,000 rupees (between around €15 and €80) depending on the size that, in turn, can range from 200 to 1,000 litres in capacity. In general, 80 per cent of the tanks purchased have a storage capacity of 500 litres.

Finally, families in many traditional dwellings continue to use traditional wells. These freely accessible wells, located away from private property, continue to be highly popular among the towns' poor inhabitants. In reality, the water provided through the poorly operating water scheme has become more of a back-up resource for users, who mainly obtain their water from less restrictive sources. Thus, not only is access to water in small towns symptomatic of the transitional nature of these agglomerations, which are both rural and urban and where there is a combination of public, private and traditional providers, but it also illustrates the planners' failure to improve operation of the service, undermining efforts to expand access despite decentralization.

The myth and reality of service decentralization

"Technical" capture rather than "elite" capture

Decentralized management of the water service, as set up in the small towns studied, has not led to any notable improvements in the service level. This significant finding, which contradicts a recent – and embellished – evaluation of water service decentralization in other Indian towns (Gupta *et al.*, 2011),[18] challenges the simplistic correlation between the geographical size of the service and managerial effectiveness (Blair, 2000). As already illustrated in other towns, the reforms can lead to the "deprofessionalization" of technical decision-making, resulting in local power struggles, enhanced patronage interests (World Bank, 2007) and even counterproductive corruption processes (Asthana, 2008), thus giving lie to the myth that the reforms automatically guarantee participation, democratic power sharing and local service improvements. Furthermore, many analyses show that, in certain cases, decentralization can strengthen the position of the local elite, which secure any benefits obtained for their own ends ("elite capture"[19]), notably by "hijacking" some of the decentralized public services, such as water supply or sanitation (Weis, 2005; Ban *et al.*, 2010), and ultimately widening the gap between the rich and poor neighbourhoods.

In the towns studied, where the population continues to be geographically segregated by caste (the Dalit areas usually being consigned to the outskirts of the town, with the town centre mainly home to the upper castes), the majority of municipal-elected officials belong to the same dominant group of local community leaders. However, it would appear that the specific nature of the water supply sector ensures that any hypothetical infrastructure capture by dominant groups is minimized. Thus, the fragmented distribution of facilities across districts seems due not to intentional discrimination, but rather to a range of technical factors. The real authority to expand the piped water service remains in the hands of its designers and those financing the network's technical infrastructure, rather than in the hands of the municipal managers, whose scope for action is still extremely limited. Although the monthly service charges are lower than those recommended by Jal Nigam (and rarely collected), work to expand access to the network remains hampered by plumbing costs, the technical capacity of the network, its electricity supply, as well as a host of other technical factors outside the scope of the local managers' remit.

The situation as regards access to water in the towns studied reveals the important role played by the supply network in determining the level of service provided, which supports similar observations seen in other towns in India, such as Delhi, Mumbai and Hyderabad. "There is also a technical dimension. In fact, connections can be given only to households in areas where this is physically possible" (Huchon and Tricot, 2008: 37). In both small towns and large cities, access to the supply network determines the level of service received: due to their proximity to the distribution system infrastructure, the historic urban centres always have a better water supply than the newest districts, which are often located outside the reach of the existing network.

As the design of this distribution system is driven by technical and demographic concerns defined by the regional para-governmental agencies, fears that the local authority may misappropriate the service for the benefit of only some parts of the population thus appear misplaced. With the exception of the installation of water pumps (which is, by its very nature, more decentralized), expansion of the water service depends mainly on a number of technical factors and so is largely independent of the municipality. More generally, these modes of access underline the major role played by the supply system in the very fabric of the town (Tarr and Konvitz, 1987; Coutard, 2005) and disproves the theory that reforms to networked services will reinforce growing urban fragmentation (Graham and Marvin, 2001). Under this setup, as the body financing the infrastructure and responsible for its transfer (via the Jal Nigam agency), it is the provincial government that appears to play the lead role in expanding access to the service.

The continuing role of the state

Comparison of the technical systems within the four towns studied, which have similar needs and are located within the same region, reveals that the regional authorities of Uttar Pradesh's government are not treating all the towns equally. Existing network infrastructure provides considerably different levels of service, with the quantity delivered to the tap varying by as much as threefold (from 70 litres per capita per day in Phulpur to 200 litres per capita per day in the tourist area of Kushinagar).

Installed in each of the towns studied during the 1970s (when the villages gained urban administrative status), just a few years later, the water supply systems were inadequate to the task of meeting the growing demand. With population growth, upgrading of the infrastructure is triggered once the size of the town reaches the critical threshold of 20,000 inhabitants. However, the actual outcome of each infrastructure resizing project is contingent upon a range of cumbersome administrative procedures which can sometimes take years to complete, as the regional authorities do not consider the town a priority. Project approval requires the agreement of an array of institutional stakeholders with sometimes competing interests. In Siddarth Nagar, for example, the engineers opted to modify the initial project rather than attempt to negotiate a possible compromise with the Department of Railways, which was refusing to allow them to lay pipework under the railway line. This administrative process is also being hampered by the increased involvement of the municipality due to the tension this creates. In practice, Jal Nigam continues to be in charge of all technical supervision and inflexibly designs each supply network infrastructure expansion using its own statistical population projections rather than basing this on the requirements submitted by the local authority, which is only actually consulted to initial the final project. As in other large towns in India, rather than working to improve the poor state of existing infrastructure, which risks gradually deteriorating further, the engineers are instead playing catch-up (Zérah, 1999).

In all the towns studied, funding for this technical infrastructure is very heavily subsidized (between 90 and 100 per cent) by the regional and/or central government, or even by outside institutions (as in Kushinagar, for instance, which received special assistance from the Japanese government to improve water supply within the tourist district, where the main Buddhist pilgrimage sites are located). Since the introduction of the decentralization reform, the municipalities have to contribute 10 per cent of the total cost of each new project. However, this change in policy, intended to foster ownership of the infrastructure by the local authority, is proving difficult to implement due to the lack of municipal resources available in Uttar Pradesh and in its small towns, in particular, where local tax revenues rarely exceed 10 per cent of the total revenue available, which is not always the case in other states, such as the Punjab (Mahadevia and Mukherjee, 2003). The vast majority of investment still comes from public aid granted by central and regional government as part of urban infrastructure renovation programmes.

One of the last programmes, launched in 2005 by the central government, was the Jawaharlal Nehru National Urban Renewal Mission (JnNURM), which was providing substantial amounts of funding for urban modernization. The majority of this financing has been allocated to 65 major cities of "national importance" (as part of the Sub-Mission for Urban Infrastructure and Governance). Only 21 per cent of the remaining budget has been made available to the 640 small towns involved in the Urban Infrastructure Development Scheme for Small and Medium Towns (UIDSSMT), despite the fact that these towns are home to 57 per cent of the urban population (and thus project beneficiaries). Therefore, although there is intense competition between these small towns to access the funding available, these municipalities rarely have the administrative skills required to "draft funding requests (DPR – Detailed Project Reports) and raise the necessary resources" (Kundu and Dibyendu, 2011). Consequently, within the same region, rather than funds being granted based on the needs of each municipality, this investment ends up being directly allocated in accordance with the political priorities of the state and the inconsistent decisions it is compelled to make.

Due the specific features of the sector, the water supply service thus remains highly dependent on the piped network infrastructure, the overall technical structure and funding of which seems to have remained unaffected by the decentralization reforms in their current form. Thus, as things stand, expansion of the water service rests first and foremost with the technical and non-decentralized decisions made by the Jal Nigam regional office and the investment decisions of the states, rather than with the municipalities.

Conclusion

There are many lessons to be drawn from the analysis of the difficulties encountered during the decentralization of water services in the small towns studied. First, without substantial investment to upgrade the water distribution networks, any reform of the service is doomed to achieve poor results.

Recent evaluations of the JnNRUM reveal that only 8.9 per cent of the infrastructure projects approved by the government since 2005 have so far been completed and highlight a range of issues, including corruption, bureaucratic delays and lack of local capacity (Supreme Audit Institution of India, 2013). Furthermore, the service is deteriorating as a direct result of the lack of any skills transfer to the municipalities. Thus, unless municipal authorities are truly involved in their technical management, infrastructure upgrading programmes will merely continue to help create a "mythical public good" (Ruet, 2002). Finally, despite these failures, the municipality's introduction of public standpipes in districts with poor network coverage illustrates the positive "day-to-day mediator" role that local elected officials have been playing since decentralization (Berenschot, 2010). Nevertheless, this increase in alternative "off-network" facilities also reflects the growing "hybridization" of the service (Coing, 1996), which, far from being specific to India,[20] risks undermining the concept of equal access for all inhabitants regardless of their means. The poor outcome of water service decentralization thus highlights how vitally important it is to address the technical needs that the municipalities, particularly those in small towns, currently still seem to have. Regardless of the reform introduced, close synergy needs to be generated among the local authorities, technical agencies and donors before any expansion of the service can be carried out.

Notes

1 In the literature on developing countries, the term *decentralization* covers a wide range of processes, from administrative devolution through to the implementation of a policy for local democracy. In this chapter, the term is understood to mean the process by which a central government formally devolves responsibilities and authority to local public authorities (Jaglin *et al.*, 2011).

2 The Census of India is carried out every ten years. The results of the 2011 Census of India will be released over the course of the next few years.

3 These towns meet the demographic and economic criteria used by the Register Central of India (RGI), which defines a town as having more than 5,000 inhabitants, a population density of at least 400 inhabitants per km^2 and at least 75 per cent of the working population that does not work in the primary sector. Every ten years during the Census, several towns are redefined as urban centres under these criteria, whereas some revert to village status.

4 In 2004, "India shining" was the slogan used by Atal Behari Vajpaye of the BJP, who stated his intention to transform the country into a major world power by 2020.

5 Thus, for the economist Amitabh Kundu (2011), the apparent service improvement is primarily a statistical mirage created by the official definition of "access to water", as, since 2001, households with a water source more than 100 metres away from their home are now included in the new, vague category of "away", created for the 2001 Census.

6 The site of Siddhartha Gautama Buddha's cremation, dated at around 483 BC, is believed to have been in Kushinagar (which literally means "place of joy"). The government of Uttar Pradesh is eager to develop the town's tourist potential; see: www.kushinagar.com.

7 First, between 2008 and 2012, four visits were conducted to each of the four towns for a period of between ten days to three weeks, during which time the author surveyed a sample of users (based on the location of the neighbourhood within the town, its age and social composition) and interviewed municipal managers (politicians, devolved civil servants, local technicians and employees) to obtain first-hand empirical data on how the water service operates. During these visits, a number of trips were also made to these towns' neighbouring provincial administrations to meet governmental managers and engineers, notably to obtain technical data on the network infrastructure (design, service capacity, means of financing, network maps, etc.). Finally, three further one-week assignments were undertaken interviewing ministers in Lucknow, the state capital. Throughout the research period, this immersion was greatly facilitated by the institutional, technical and financial support of the *Centre de sciences humaines* (CSH) in New Delhi, to whom the author would like to express his sincere appreciation.

8 The Uttar Pradesh Municipality Act 1916 was amended in 1994, from Clause 224 through to Clause 235-A, to include the legislative changes introduced by the seventy-fourth amendment to the Constitution of 1992.

9 This is not the case in rural areas, where Jal Nigam continues to be fully responsible for the entire drinking water service when this is provided through a distribution network.

10 A caste system remains in Indian society, which subdivides the population into hierarchical social strata. In this case, the engineers working for Jal Nigam are predominantly from the upper castes, whereas the majority of the elected officials in the small towns studied belong to castes considered inferior under Sanskrit tradition.

11 Interview at the Jal Nigam offices in Mughal Sarai, August 2008.

12 "Muslims don't take baths" was what the (Hindu) engineer for Siddarth Nagar said during an interview on 29 September 2009.

13 *Dalit* is a term used to describe the group of people previously regarded as untouchable (also called *harijans* by Gandhi), who are excluded from the caste system in India and very often relegated to carrying out tasks regarded as ritually impure. For more information on "untouchables" see Deliège (1995).

14 Interview in Allahabad, Jal Nigam, 2009.

15 Although not followed in practice, these official recommendations to charge for the use of standpipes reflect a willingness to introduce a cost recovery policy, thus breaking away from the policy of free access generally seen in Indian towns.

16 This sample was taken from different neighbourhoods within the town. These neighbourhoods were selected for their geographical location (on the outskirts, in the centre) and economic factors (a neighbourhood with a high Dalit population and a neighbourhood with a low Dalit population).

17 It is not possible to put a figure on the number of cases of waterborne diseases seen in the towns studied, as the district hospitals keep no precise records of these.

18 This study examines 27 of the 63 towns selected to benefit from the JnNURM (Jawaharlal Nehru National Urban Renewal Mission) and concludes that greater autonomy improves the performance of the service provided.

19 In Indian literature, the term "elite capture" usually refers to the process by which local leaders use the benefits of decentralization to further their own ends (see e.g. Kundu (2009), whereas in Africa it usually refers to the central government's hold over the local elite.

20 Initially introduced by the World Bank on Nigeria (Lee nd Anas, 1989), then extended to cover other Southern countries, there is now a wealth of literature on the (private, community-based) alternatives available for replacing or supplementing the public water supply service. See specifically on India Llorente and Zérah (2005).

10 Water disputes in Soweto

Inequalities, social mobilizations and recourse to the law

Julie Aubriot

> Everyone has the right to have access to sufficient water.
>
> (South African Constitution, Bill of Rights, Article 27, 1996)

Since democracy came to South Africa in 1994, guaranteeing the right of access to water has been a government priority. But while the country has one of the most robust constitutions in the world in terms of social rights, persistent inequalities remain (Seekings, 2007; Terreblanch, 2004). In particular, the constitutional ideal of universal right to water is still an idealistic goal for millions of citizens for whom access to economic and social rights cannot be taken for granted.

The South African government has made substantial investments in the water sector, notably a "Free Basic Water Policy", in terms of which 6 kilolitres of water are provided monthly for free to all households. However, the quality of the service remains a source of controversy and there is a strong sense of discontent among poor urban populations in this regard (Eales and Smith, 2011). In the wake of debates about the concept of the "right to the city", access to urban services is increasingly thought of in terms of rights (Parnell and Pieterse, 2010; Parnell and Boulle, 2008), particularly in the water sector (Bond, 2010). A number of social groups are active with regard to the question (Ballard *et al.*, 2006), and have mobilized around social rights issues, either symbolically or in courts. Such groups have mainly attempted to address the issue of access to fundamental goods and to the city. Consequently, local governments have dealt with many recent legal cases about the right to health services, accommodation and water. Furthermore, these demands are being made in a tense political context involving social movements protesting the political approaches taken by successive ANC governments and the authorities responsible for implementing public policy on essential services and applying an approach largely informed by concerns over cost recovery.

This climate of tension is particularly palpable in the City of Johannesburg which, faced with the challenge of introducing the principle of cost recovery, while simultaneously encouraging universal access to quality infrastructures, implemented the Gcin Amanzi Project (GAP) in Soweto, the city's largest township. The Gcin Amanzi Project, based on a major management contract with a private sector company that was tasked with installing prepayment meters on a massive scale, was the object of vigorous criticism from a number of social movements which, after several years of street protests, switched their focus to the courts. In the Mazibuko case (2006–2009), these movements took on the municipality of Johannesburg and its water provider, Johannesburg Water. Activists were able to question social policy on water and criticize the massive use of a controversial socio-technical tool, namely prepaid water meters.

But beyond technical and economic issues, an examination of social relations and the political power relations in which they are inscribed provides a valuable perspective from which to analyse the capacity of a specific group of consumers, namely the residents of poor urban areas, to develop collective resistance strategies characterized by varying degrees of organization. In order to defend a vision of fair and acceptable access to water, the approach developed was initially based on a range of civil disobedience tactics, and later on a recourse to the law.

After having recalled the context in which the Gcin Amanzi Project emerged, we will analyse the conditions of the politicization of the crisis around access to basic services and explore how the recourse to the law has transformed the strategy of protesters and made injustice visible. We will demonstrate how the process, despite a legal defeat in the constitutional court, has led to significant changes in social policy on access to essential services for marginalized urban populations.

The Gcin Amanzi Project

The legacy of non-egalitarian services and the reform of the institutional framework of water management

In the mid-1990s, Johannesburg was still organized on the basis of the separate development policy applied during the apartheid regime. Re-equilibrating urban areas thus implied a homogenization of access to essential services, especially in the townships which were dormitory towns on the outskirts of the city populated by non-whites and underequipped in basic infrastructures. Moreover, infrastructure development was based on racial segregation.[1] Within the city of Johannesburg different access to water services were coexisting, ranging from direct connection to a quality network to no connection at all. In 2000, the population of Johannesburg was estimated at three million people. At the time, one-third of this population, who lived on an average to high level of income, had access to water via individual connections equipped with post-payment meters. Another one-third, most of them resident in the township of Soweto, lived on a low level of income and were provided with water without meters for a flat rate based on a consumption of 20 m^3 per month at 125 ZAR.[2] The last one-third of the population was made up primarily of poor people living in unofficial neighbourhoods and receiving a minimum level of service for free via tanks or communal standpipes. Transforming these unequal services into a universal system of water provision at the municipal level involved extending the network to cover areas that were not connected. This also meant improving existing poor-quality networks, especially in the townships.

After apartheid, cities previously divided into white municipalities and black townships were reorganized across the country with a view to ensuring that local authorities operated in the same conditions, thereby guaranteeing equal treatment for rich and poor in terms of public services (Jaglin, 2003b). The City of Johannesburg was set up by uniting the 11 former municipalities under the umbrella of a single metropolitan authority; however, Johannesburg's financial position continued to worsen, a process that culminated in bankruptcy in 1997. Supported by the World Bank, the "Igolo 2002" plan was introduced in 1999 with the objective of transforming Johannesburg into a "world-class city". The plan was intended to mirror at the municipal level the cost recovery policy adopted as part of a wide-ranging programme of national development (the "Growth, Employment and Redistribution Plan"). Furthermore, the implementation of a service reform programme led the municipality to set up a private company – Johannesburg Water – owned 100 per cent by the city. The company's mission was to guarantee universal access to water while reducing physical and commercial losses at the municipal level. In 2001, loss through leaks and other causes was estimated at 43 per cent. In the same year, Johannesburg Water signed a five-year management contract with a foreign private company, a subsidiary of Suez. This contract was motivated by a number of elements, notably including making improvements to the service, transforming the company into a financially viable entity in the long term, and incorporating six previously independent entities into a single body with a sole modus operandi. While this management contract was considered internationally as an example of a "successful" public–private partnership (Marin, 2009), it was also subject to a great deal of protest, notably on the part of residents, unions and social movements, which regarded it as a "privatization contract".

Transforming the urban poor into "responsible" citizens

It was against this backdrop that Soweto, Johannesburg's oldest and largest township, was identified as the cause of the financial deficit in the local water sector. In September 2003, Johannesburg's mayor, Amos Masondo, announced the launch of a vast operation designed to renovate pipelines and homogenize water services in the city. The Gcin Amanzi Project ("gcin amanzi" means "water conservation" in Zulu) involved the installation of 170,000 prepaid meters, which represented approximately, 32 per cent of connections in the city, and 67.5 per cent of connections in the townships (Marin *et al.*, 2009). The Gcin Amanzi Project was at once the largest pipeline renovation programme ever undertaken in South Africa, and Johannesburg's most ambitious approach to managing its urban water provision needs. The programme, the initial cost of which was estimated at around €49.4 million, was, at the time, the biggest meter installation scheme in the world. If successful, the municipality would make a highly positive impact at the international stage. The municipality argued that there was less water loss in areas with meters, or, in other words, in areas inhabited by residents with average and above-average income, than in areas without meters in which inhabitants paid a flat rate (see Table 10.1). The authorities offered a number of explanations for this situation, including the existence of a "non-payment culture", a high rate of consumption (approximately 66 m^3 of water per household per month), and the fact that the pipeline network was in a parlous state. Given the context, the municipality and the operator were quick to make the townships a priority. In 2004, a pilot project was set up in Phiri, one of the poorest neighbourhoods in the city.[3] Work on the project continued apace up until 2007.

As well as renovating the public water pipe network, the project involved a number of different missions. Households targeted by the Gcin Amanzi Project were able to benefit from leaks being fixed in their properties and having their water debts written off,[4] but this was dependent on their acceptance of the prepaid metering system. Households refusing to take part in the project did not have their debts written off and were forced to accept water supplied via standpipes and water tankers. If, under such circumstances, water was provided free of charge, the amount delivered was limited to between 6 m^3 and 10 m^3 to keep consumption down. In other words, households that refused to accept prepaid meters were disconnected from the water distribution network.

The project also had an important social component in that, for the municipality of Johannesburg, the objective was to change the urban poor's attitude towards paying their water bills. During the apartheid period not paying water bills was a political gesture designed to destabilize black local authorities and "render the townships ungovernable" for the ANC. In effect, black local

Table 10.1 Estimated physical and commercial losses in Johannesburg in 2004

	Physical losses (%)	*Commercial losses (%)*	*"Non-revenue water": total losses (%)*
Areas with meters	9	10	19
Areas without meters	11	58	69

Source: Johannesburg Water (2004).

authorities had to fund themselves by introducing urban taxes payable by township residents, and this at a time when white townships were able to rely on taxes from industry and business to fund their budgetary needs. Local urban services in general and water services in particular were a central factor in constructing political legitimacy at both the central and local levels since the beginning of apartheid.

With the advent of democracy and the end of the apartheid era, this situation was no longer acceptable in South African municipalities. In 1994, the payment of services was a central aspect of the strategy for developing the nation and an important part of efforts to "normalize" relations between the state and its citizens. In effect, the government took the view that democracy not only provided rights, but also implied responsibilities, specifically in regard to the payment of services. However, in spite of large-scale campaigns designed to promote awareness among members of the public, advocates of the principle of cost recovery were soon confronted by poverty and, in some cases, by unwillingness to pay.[5] Prepayment meters were introduced as a technical solution to the problem of non-payment, thereby leading to a form of coercion in a number of economically deprived areas. For the municipalities, this new technology made it possible to supply every household with free water, with residents obliged to purchase additional water using a prepayment card. Consequently, prepayment meters, installed exclusively in areas historically inhabited by black people on low incomes, mirrored urban segregation borders inherited from the past. Under these conditions, the obligation to pay for water was perceived as an additional injustice forced upon low-income users (McDonald and Pape, 2002).

By the end of the pilot phase, Johannesburg Water was able to boast that less water was being wasted with an estimated return on investment of 36 per cent and the rehabilitation of 116 km of pipes. In addition, nearly 100,000 prepayment meters had been installed by this time. But if the development of a differentiated level of service and of specific socio-technical tools can be seen as a response to the idea that "the ideal model of public services based on the expansion of universal services by means of a Weberian administration … ha[d] revealed its limits" (Jaglin and Zérah, 2010: 8), in reality, taking into account the socio-economic specificities of the area targeted, this differentiation was perceived by water consumers not as an approach designed to facilitate access to services, but rather as a way of perpetuating inequalities. For the municipality and the water company, the project – initially unpopular with consumers – was more of a socio-economic and political challenge than a technological or financial one (see Table 10.2).

Table 10.2 Statistical impact of the Gcin Amanzi Project

Consumption prior to the project	66 m^3 per property per month
Consumption after introduction of project	12 m^3 per property per month
Percentage reduction in water use	81
Percentage of users purchasing no water in addition to the free water provision (6 m^3)	48
Monthly cost of water	42.05 ZAR
Amount invoiced monthly prior to the project (flat rate)	149 ZAR
Total reduction in terms of water purchased from Rand Water (since July 2004)	Savings of 98,000,000 m^3 of water at a cost price of 290 million ZAR

Source: adapted from Johannesburg Water (2010).

A politicization process in two phases

Social mobilizations and civil disobedience

From the outset, the project aroused a series of collective protest movements characterized by acts of civil disobedience, vandalism and more generally, "illegal activities" (Mouchard, 2003). In August 2003, local people set up the Phiri Concerned Residents Forum (PCRF) with the support of the Coalition Against Water Privatization (CAWP) and the Anti-Privatization Forum (APF), one of the influential "new" social movements in Johannesburg (White, 2007; Gentle, 2008; Tournadre-Plancq, 2008).

Social movements came in all shapes and sizes. Ordinary militants, most of whom were unemployed or retired (for example, the "grannies" of Soweto[6]) protested alongside the leaders of social movements who, often living outside the townships, had roots in the middle and upper classes, and as a result, these upper echelon groups had a greater degree of social and intellectual capital than did the base of the organization.

These protests, first individual then collective, were at once procedural and technical. They focused on criticizing the methods employed by the municipality, which they characterized as a lack of consultation, restrictions on consumption, an obligation to pay for water (via the introduction of meters) and an insufficient quantity of free water for poor households (these households often included more than eight people, the number of individuals on which the free water policy was based). Protest took the form of *Vul Amanzi* ("more water" in Zulu), whose slogans, including "Break the Meters, Enjoy the Water", "Our Water is not for Sale" and "Phansi Privatization! Phansi" ("Down with Privatization") illustrated the strength of opposition to the "privatization" of water and the principle of cost recovery. Tactics included demonstrations, petitions, weekly information and planning meetings, filling up channels dug during renovation works, cutting pipes installed by Johannesburg Water, and sabotaging prepaid meters.

These campaigns, which consisted in a number of actions ranging from sabotage to demonstrations and meetings, served to make the cause visible. Furthermore, the illegal activities were "sectorial" (Péchu, 1996) in that they had a defined "local" objective, focusing on areas directly affected by the problems identified and causes defended. Meters were torn down in local neighbourhoods and marches on the water company and municipal offices were organized, since it was argued that both organizations were responsible for the injustice denounced by residents. This "politicisation of day-to-day life" (Mouchard, 2002) mirrored a routinization of illegal practices, a series of militant acts that served as a political response to an immediate social need. While it is fairly difficult to estimate the scope of the protests, according to the water company, between 40 and 50 per cent of prepaid meters in Soweto were illegally disconnected, with 4,000 meters torn from the walls.

The "privatization" of services and a certain idea of justice as catalysts of protest

While protests focused on a number of different technical and procedural aspects of the Gcin Amanzi Project, they also reflected a more generalized discontent with the political stance of the ANC government that had come to power in 1994. In effect, while giving priority to the vital problem of access to essential services, social movements called into question socio-economic relations as a whole, defining themselves as "anti-capitalists", "anti-ANC" and "socialist" (Ballard *et al.*, 2006). The struggle against "privatization" was an important factor in protests against the project. Even if they were politically heterogeneous, militant organizations were supported by unions, and by intellectuals, either in or close to social movements (Bond, 2000; McKinley, 2004), who all shared a radical, anti-privatization and anti-cost recovery ideology that had its roots in a critique of reforms introduced at the municipal and national levels, which they accused of belonging to a "neo-liberal programme". The objective was not only to reconnect or disconnect prepaid meters, but above all to organize an ideological campaign against the ANC and its "neoliberal" policies. In effect, militants believed that the ANC had betrayed them by falling in with the capitalist regime and forgetting the social promises made when it came to power in 1994.

As a counterpoint to their political activities, organizations militated in favour of free access to essential services, applying the concept of water as a public good and pointing out that access to the resource was a constitutional right. In effect, as well as calling into question the concept of public services, social movements and government agencies were characterized by differences in their approaches to water. The social movements considered water, housing and electricity as public goods that should be managed exclusively by the public sector. However, for the authorities, water was a commodity, i.e. a service with associated costs in terms of its use and distribution. While rejecting the principle of cost recovery, the involvement of private foreign operators and the installation of prepaid meters, the social movements also invoked the right of access to water entrenched in section 27 of the Constitution.

Social movements justified sabotaging prepaid meters on the grounds that it was "fair" and legitimate to disconnect them in order to further the cause of the right to water. Civil disobedience was thus a way of protecting rights already recognized by the Constitution. In effect, the social movements were attempting to "found the kind of society that should have been created by the 1994 national revolution" (Tournadre-Plancq, 2008: 155). Although their perception of social rights was sometimes partial, superficial and even erroneous, it nevertheless provided a repertoire of justifications and a source of legitimization for protesters.

The end of street protests and the beginning of a legal strategy

Against a backdrop of arrests, intimidation and threats from the authorities, collective militant resistance gradually ran out of steam (CAWP and APF, 2007). The social movements' inability to influence the policies of the city government led to a change in emphasis (Aubriot, 2012). As a militant explained: "We tried in every way we could to negotiate with the City. But the municipality laughed in our faces and held to its line, according to which meters are the way forward. Now, we're fighting them in the courts".[7] In 2006, social dialogue and local politics were replaced by a legal approach that gave the protest movement a second wind. A more complex network of actors emerged in support of militant organizations, which tended to consider the law as a tool of oppression rather than as an instrument of general mobilization. After three years of campaigning, five Phiri residents, supported by social movements involved in mobilizations from the outset, as well as by a human rights group and a renowned constitutional lawyer, filed a lawsuit against the City of Johannesburg, Johannesburg Water, and the national minister for Water Affairs and Forestry: the Mazibuko court case – Mazibuko being the name of the first applicant. Protest was undergoing a process of juridicization. Within this framework, the approaches and justifications applied by the social movements changed almost beyond recognition. They no longer indulged in a uniquely rhetorical usage of what might be termed "subjective" law, but instead employed the positive law sanctioned by the state. This approach to the law enabled protest movements to emerge from a systematic politicization of the debate and to focus instead on the technical and social aspects of the Gcin Amanzi Project and the city's water policy.

The grievance presented to the South Gauteng High Court of Justice was based on the fundamental right to water enshrined in the Constitution (1996). It requested South African courts of law to rule on the content of a constitutional social law and on programmes based on a specific interpretation of that content. The grievance addressed a number of aspects of the municipal water policy. First, according to the applicants, prepaid meters restricted their right to water by automatically limiting access without prior notification. Second, they claimed that, considering the number of people making up poor households, the volume of free water provided ($6\,m^3$ per household per month) was insufficient. The applicants drew attention to the fact that while in line with the recommendations of the World Health Organization, the Free Basic Water Policy (FBWP) was calculated to meet the essential needs of eight people per household at 25 litres per person per day, the policy was deeply flawed in that, on average, households in Soweto comprised 16 people. Third, the applicants called into question the method used to target social grants, highlighting the fact that free water was provided on the basis of housing units, a notion intimately linked to the idea that one property corresponded to no more than a single household. However, in reality, the free water delivered was used by all those living in the property, no matter how many of them there were (different families, tenants, etc.). Finally, the applicants argued that Johannesburg's water policy was discriminatory in that prepaid meters were being installed exclusively in Soweto, a part of the city characterized by precarious socio-economic conditions, and that the policy forced poor households to pay for their water in advance at a time when wealthier households continued to enjoy the benefits of a traditional invoicing system. The court case triggered a re-emergence of the question of state discrimination and the issue of a confrontation between a part of the population that was poor and black, and another section that was better off and white. Sensitive questions thus took centre stage in a country in which democracy is still under construction, political, societal, economic and social transformation is partial and selective (Vircoulon, 2004), and, at the same time, there are growing concerns about South Africa becoming a country divided not along racial lines, but rather according to social class (Bond, 2004).

Law as a paradoxical tool for fighting injustice

The discriminatory character of the municipal policy in question

The Mazibuko case lasted for four years and went through the high court, the supreme court of appeal and the constitutional court (Aubriot, 2012). On 30 April 2008, the High Court's ruling validated all the requests made by the applicants and went through the high court, the supreme court of appeal and the constitutional court and ordered Johannesburg Water and the city of Johannesburg to increase the volume of free water supplied and to offer a choice between prepaid and traditional meters. However, four years later, the Constitutional Court ruled entirely in favour of the municipality, rejecting the applicants' case out of hand. But what could be seen as an outright victory for the municipality and the operator was more ambiguous that it may at first have seemed. In spite of the applicants' defeat at the hands of the Constitutional Court, recourse to law made it possible to influence municipality's policy, rendering it, at least on paper, more "pro-poor". Indeed, both the operator and the municipality took almost all the arguments made by the applicants into account. Although, from the point of view of the militants, the Mazibuko case ended in defeat, it nevertheless obliged the operator and the city to reappraise their approaches to poor people's access to water. In effect, representatives of the municipality and the water company unanimously affirmed that the case had a positive impact, since it forced them to cast a more critical eye over water policy, and more generally to re-examine approaches to integrating the urban poor.

During the case, the applicants cast light on the discriminatory character of the Gcin Amanzi Project, which exclusively targeted historically poor areas inhabited by black people. Their accusations were based on the central notions of equality and non-discrimination, both founding values of the 1996 Constitution. The lawyer representing the residents of Phiri explained in an article published in the *Mail and Guardian* that "there was discrimination in the way the city treated the rich, mostly white residents of Johannesburg and the poor, mostly black residents.... Clearly it is a case of indirect discrimination based on race".[8] In response to these critiques, the City and Johannesburg Water used techno-environmental arguments to defend their decision to exclusively target Soweto and highlighted their desire to conserve water and manage demand for the resource. Two kinds of argument were deployed during the court case. The first, which was social and ethical in nature, focused on the notions of equality and non-discrimination, while the second emphasized environmental concerns, the scope of which went far beyond Soweto. Arguments of the latter kind formed the basis of the Constitutional Court's ruling that the Gcin Amanzi Project was not discriminatory in that it had been implemented, on the one hand, for environmental reasons, and on the other, because the municipality had introduced a system of social grants to help the people of Soweto.

Beyond the discriminatory aspect of the Gcin Amanzi Project, the complaint presented to the Supreme Court by the applicants reopened wounds that, officially at least, belonged to the past (Vircoulon, 2004), including racism, apartheid, discrimination, and the opposition between the rich and the poor, and between whites and blacks. The following extract is enlightening in this regard:

> inequality and discrimination sometimes derive from a paternalistic, rather than a spiteful attitude. The municipality's policy is based on the idea that the residents of Parktown have the right choose the best way of paying their bills, while it's the municipality itself that decides what's good for the residents of Phiri. This is an insidious form of racism of the kind that was common during the colonization and apartheid period, and which, even if it is based on good intentions, remains offensive.[9]

The City took these accusations very seriously, even though the Constitutional Court had rejected them. In the young democracy of South Africa, municipalities cannot leave such accusations unaddressed or allow the terms of the debate to be defined publicly in this way. In order to silence those critiques, the City of Johannesburg suggested a more egalitarian a priori technical solution involving the installation of prepaid meters throughout the city for both water and electricity. However, and perhaps because of the high costs involved, the idea has not yet been politically validated.

Inclusivity and the "invisible poor": the effects of the Mazibuko case

From the beginning of the project, prepaid meters crystallized tensions. After the Constitutional Court ruled that they were legal and constitutional, they remained a vital part of Gcin Amanzi. However, following the court case, meters were the object of technical improvements designed to take the poorest households into account. The new meters provided a minimum of 40 litres per hour at low pressure – the so-called "lifeline flow" – over and above free water already supplied, thus giving consumers continuous access to a minimal volume of water. From the City's point of view, the approach solved the problem of automatic disconnections raised by the applicants during the court case. A minimum level of security in the form of 1,000 litres of "emergency water" could also be triggered four times a year in exceptional circumstances. In addition, 2,000 litres of additional water could, again in exceptional circumstances, be allocated in case of "special needs" at the request of the local authorities. Nevertheless, according to municipal services, the system has not been very efficient in that residents are not correctly informed about how to use it, or even that it exists. Furthermore, the social acceptability of these tools remains a moot point to the degree that, unlike in the initial phase of the project in which residents were able to refuse to have prepaid meters installed and, in lieu, receive an inferior service, prepaid meters are remaining obligatory.

In order to encourage households to accept the new generation of prepaid meters, the municipality has introduced social grants. However, in spite of this attempt to transform prepaid meters from objects of controversy into tools at the service of a wider social and environmental policy, there was still risk of rejection.

The influence of the Mazibuko case on the evolution of social aid

In South Africa, poverty remains an issue of central importance. Even if it is far from wealthy, the South African state has the capacity to implement social measures to combat poverty, notably through social grants calculated on the basis of the level of income of vulnerable groups (handicapped people, pensioners, children, etc. For most poor households, they represent an important source of revenue (Mubangizi and Mubangizi, 2005; Triegaardt, 2007). The policy to provide free water at the national scale, a universal policy that does not target specific groups, is the cornerstone of a political desire to guarantee water for all. In 1998, the City of Johannesburg introduced specific social grants[10] subject to revisions, with notably, in 2009, the "Expanded Social Package" (or "Siyasizana"), inspired by the Mazibuko case. Not exclusively dedicated to water issues, this approach proposes grants for sanitation, waste management, electricity, taxes, rent and job-seeking allocations. The system, designed to address the multi-dimensional nature of poverty more effectively, introduced a new way of calculating poverty levels. All people with a monthly income of fewer than 3,660 rand were able to benefit from the Expanded Social Package. Focusing on the provision of access to water at advantageous prices, the system enabled people who had declared themselves indigent to receive, free of charge, a volume of water of 10, 12 or 15 m^3 depending on their level of poverty.

At first sight, this new approach seemed satisfactorily to take into account the arguments put forward by the applicants in the Mazibuko case. On the one hand, the monthly allocation of free water was increased in line with levels of poverty. On the other, the volumes proposed seemed to suggest that the poorest households would spend nothing on water and that, even if additional water had to be purchased, it would be reasonably priced. Nevertheless, a more critical interpretation of these evolutions reveals that what interested the municipality was to bring two major debates – the first on the minimum amount of free water to be provided, the second on the capacity of poor households to pay their bills – to a definitive close.

The new social grants system included a number of original aspects, among them a new way of calculating poverty, making it possible to allocate different levels of aid and to define access to social grants for tenants or people living informally in "backyard shacks" or in accommodation shared by a number of different families. This change was of central importance in the evolution of the system. Historically, relations between individuals and the municipality were conditioned by the payment of services and, therefore, by the existence of an account for every household unit enabling the municipality to send out invoices and for residents, in turn, to pay for services and make tax contributions. Prior to the introduction of the Expanded Social Package in 2009, social grants were delivered via municipal accounts, thereby evicting, in a de facto manner, formal and informal tenants, of whom there were many in the townships.

Consequently, the major new element introduced by this system was the targeting, not of household units via individual accounts managed by the municipality, but of individuals themselves, the objective being to reduce the number of targeting errors and exclusions. This important nuance provides a clearer vision of the composition of household units, and consequently makes it easier to target grants more effectively in function of how many people are living in the properties declared to the municipality.

These evolutions show that the City was aware that the notion of the household unit used initially to manage the free provision of water was inaccurate in that single properties contained more than one family. While attention had already been drawn to this phenomenon in other municipalities, such as Cape Town (Smith, 2004), in Soweto it was highlighted by the Mazibuko court case. The municipality took this aspect into account in allocating social grants on the basis of individual needs rather than on the notion of property, an approach that made it possible to address the issue of the "invisible poor" (Parnell, 2005). In spite of these notable evolutions, early assessments of the system revealed that only 10 to 20 per cent of the people in Soweto entitled to social grants actually received them. In January 2010, only 40,000 people were registered, notably due to the fact that registration sites had not been sufficiently developed. Finally, phenomena of exclusion persisted: over a million people living in "informal" residential areas were still unable to benefit from social grants for water.

The study of collective protest against inequality in regard to access to water, both on the streets and in courts of law, provides an opportunity to place the question of power relations between stakeholders, and of their politicization, at the heart of the debate about essential urban services. The Mazibuko case and the protests that led to it demonstrate the emergence of consumers capable of applying a variety of approaches to contesting the modus operandi of water companies and of influencing policy by demanding a greater degree of transparency and justification in regard to technological strategies and the kinds of socio-technical systems deployed. In a broader sense, this serves as a reminder that the principal problem in regard to the provision of essential services is political, and that, in numerous contexts in developing or emerging countries, the issue of access to water raises the question of the place of poor people in the urban environment (Coing, 2010). In order to construct the services of tomorrow, a "shared diagnostic" must be developed.

Conclusion

The Mazibuko case is a prime example of how debates on urban services for the poor are embedded in wider political divisions, not only at the urban but also at the national level, in a context of persistent social and economic inequalities. More generally, it shows that it is not possible to keep urban services separate from local political dynamics. Furthermore, the case demonstrates that the law can be used militantly to influence public policy. In spite of the limits of the legal approach, it has obliged agents of the city government, of the water company and, to a certain degree, political "decision-makers", to take a critical look at their approach to running the city. After several years of protest more likely to be criticized than heeded by the municipal government, the adoption, on the part of residents, of the language of the state enabled them to have their voices heard and understood by the authorities, who were forced to respond to their demands. This confirmed that the

> law finally provided the possibility of acting against the state. In effect, the language of the law provides, beyond the strictly judicial sphere, a

> way of expressing demands in a manner that can be understood by state authorities and is likely to have a real effect on their public policies.
>
> (Agrikoliansky, 2010: 233)

A variety of protest campaigns characterized by various approaches, carried out both on the streets and in courts of law, have made the development of public policy a somewhat tormented affair. However, these kinds of disputes can create the sorts of preconditions required for fairer policies. Finally, the right to water, initially a symbolic instrument of contestation deployed in "extra-institutional" spheres dominated by protest actions (Aubriot and Moretto, 2013), has gradually become a tool for the co-production of urban services in institutional spheres.

Notes

1 "In 1995, it is estimated that 97% of whites have access to water via a household connection versus 67% of Africans" (Beall *et al.*, 2000).
2 125 ZAR are worth approximately €8.78 (exchange rate: 1 ZAR = €0.0702950).
3 According to a report published in 2008 (*Johannesburg Poverty and Livelihoods Study*), Phiri is the tenth poorest of Johannesburg's 109 wards.
4 In Soweto, less than 10 per cent of the water company's bills were paid. In these conditions, the municipality expected to use the project to write off 1.3 ZAR of debt.
5 It was estimated at 13 per cent in Soweto in 2004 (Smith, 2006).
6 In contexts of high unemployment, pensioners are often the only people who, thanks to state pensions and social aid, have a regular income, however modest, and who are therefore faced with problems associated with paying for services. Furthermore, our surveys reveal that most militants interrogated had been active during the apartheid period, often with the ANC. Therefore, for them, being militants in social movements was a prolongation of rather than a premise for their militantism.
7 "Soweto starts its water war". *Mail and Guardian*, 24 July 2006.
8 "Residents were bullied into accepting prepaid meters". *Mail and Guardian*, 3 December 2007.
9 Supreme Court of Appeal, *Mazibuko vs. City of Johannesburg and Others, Head of Arguments, Respondents' Submissions*. Case 489/2008, art. 246.
10 For further background on the period 1998 to 2006, see *Mazibuko vs. City of Johannesburg and Others, Answering Affidavit in the High Court of South Africa*. Seedat (RA), paras 26–31, 2007.

Part IV

Conclusion

11 How socio-technical systems and their operators work

Dominique Lorrain and Franck Poupeau

One global reality raises serious questions. Despite all the international undertakings, some 2.1 billion people (i.e. around 30 per cent of the world's population) drink "dangerous" water that presents health risks.[1] In consequence, managing to supply everyone with this essential good is not just a matter of friendly discussion between specialists, but challenges the foundations of collective action and water policies. How do we explain why, in a world where billions of dollars are invested in the entertainment or armament industries, it is not possible to develop (and finance) programmes that will bring improvements to countries where this essential good is lacking?

A central hypothesis of this book is that in order to understand the problems of the water sector, the obstacles preventing its proper functioning and the inequalities that arise from them, we need to move away from debates about the legal status of the water operators – public versus private – and consider their actual practices. What do the protagonists of the water sector actually do? What actions do they undertake? Under what institutional arrangements? Both economic history and research on firms show that some companies are driven by profit, but others innovate and bring progress for a "fair" return. There may be public bureaucracies overstaffed due to patronage and clientilist practices, but also public enterprises that are competent and efficient. Ultimately, the results are explained by internal properties (professionalism) and the capacity of the protagonists to establish institutions that provide incentives and points of equilibrium between the interests concerned. In other words, one needs to focus on the practices, the frameworks in which they are embedded and the conditions of their implementation.

In order to develop this line of research, we begin with two cases which, in many respects, stand at the antipodes of economic *doxa*: the Chongqing joint venture and the Salt River Project in Arizona. In China, the sector remains dominated by public operators, whereas in the USA one expects to find private companies; and yet both of these cases are structured around partnerships between public and private actors. The book then goes on to present four cases which, in different ways, are seen as exemplars of the theme of public–private conflict (Buenos-Aires, Paris, Berlin, Nagpur), but which prove to be much more complex in their logic. A final section is dedicated to small technical systems and to the conditions under which they develop. The aim is to see, on the one hand, what they contribute that the existing system was unable to do, and on the other, whether they offer lasting solutions.

The cases studied teach us that, for progress to take place, the right skills need to be available. Although the water sector is not one of the new technological industries, it demands expertise at multiple levels, even in small decentralized systems, and the responses provided make all the difference after a few years. The overstaffing in Buenos Aires' public company or, conversely, the results of the La Paz–El Alto cooperatives or of the small technical systems in Maputo and Ouagadougou are testament to this often forgotten factor. However, for operators – whether public or private – to be able to apply their skills, they need a legal framework and a minimum level of trust, so that they can tackle the problems and adapt their policies to the context; the urban environments in which they operate are always characterized by change, be it in demographics, the economic base or spatial characteristics. The difficulty of the exercise, moreover, is that it takes place in a situation of monopoly, which reinforces the asymmetry in the skills between firms and municipal authorities. Unless it is assumed that the operators are naturally virtuous, trust – which is necessary to action – has to be managed. This problem of rules or the regulation of the relations between public authorities and operators is at the heart of performance. The balance sought is fragile: an insufficiency of rules can encourage opportunistic behaviour, while too many can hamper action and lead to crises.

This chapter will present three main outcomes of the combinatory perspective developed throughout the case studies. First, the articulation between these different factors contributes to the development of water policies and the approaches employed by operators – what we refer to as "water regimes". Second, it sheds light on the origins and characteristics of controversies and conflicts about water; while it is true that culture (or informal institutions) explains a great deal in terms of the spontaneous acceptance or rejection of various systems, it is equally important to deconstruct institutions in order to reveal "second-rank institutions" that explain the evolution of controversies: the role of contracts, calculation mechanisms, types of report; spheres of expression, coalitions. Third, with regard to the much-discussed subject of the possible "pathways to progress" for developing countries and the choice between large and small technical systems, the cases studied cover a wide range of context-dependent solutions.

Several combinations of factors: first results

A global analysis reveals that while some institutional choices have worked (Chongqing, the Salt River Project), others have been terminated independently of their results (Berlin, Buenos Aires, Paris, La Paz–El Alto), whereas others have had mitigated results (Nagpur). In seeking to explain these outcomes, should we attribute them to the behaviour of the operator, the attitudes of consumers or to thoroughgoing changes made to initial parameters? In answering these questions, one observation is that the same categories are to be found in all the cases. Namely we can distinguish:

1 *formal institutions*, or the rules of the game (North, 1990);
2 *second-rank institutions* that define technical standards, calculating tariffs, revision procedures, accounting norms, monitoring formats, information policies, etc. (Lorrain, 2008a);[2]
3 *informal institutions* characterized by beliefs, local technical cultures, political values and the recognition of the legitimacy of the powers and expertise of the state – all parameters to which the social sciences often refer as external "contexts" (Sabatier and Jenkins, 1993) but which play a role in structuring markets and their operations (Bourdieu and Christin, 1990);
4 *socio-technical* systems, including not only urban infrastructure, but also industrial policies designed to maintain and extend it (Tarr and Konvitz, 1987; Hughes, 1983).

One of the main results of the research presented here is that the organization of a water regime differs depending on the combination of factors. In fact, from one case to another, the balance between these three categories – the formal institutions, the "small" institutions and the informal rules – differs from one water regime to another. What is distinctive about the water sector compared with other socio-technical systems is that it is embedded in social practices that bypass the industrial strategies of the "right product" and the "right price". In some countries, as this book shows, it takes on community forms and is subject to local controls, which explains the emotional charge still associated with the status of the distribution companies. To account for these phenomena, the *connection between social practices and second rank institutions* helps elucidate the way in which water policies are – more than implemented – appropriated and/or contested. In each case, the institutional dimensions are combined in a specific way, entailing particular competences; this combination can be the outcome of public policy or, on the contrary, the mismatch between industrial policies and local practices can give rise to conflicts emanating from the community forms. The relations between these institutional and social levels thus constitute what defines each water regime. A number of conclusions may be drawn from this approach and the different cases. The introduction of a water service does not consist solely in installing a socio-technical system or in managing the power relations that govern its funding. Project results may be explained by a combination of factors that vary according to context, the kinds of problems to be addressed, the balance of political power, the organizations and sometimes the individuals involved, which may play an important role in the provision of local services of this kind, as is illustrated in the example of Uttar Pradesh and of local politicians in Berlin. There is no single approach, in the sense of a "model" of best practice, such as those recommended by the World Bank, that must be implemented at all costs.[3]

The cases studied illustrate this diversity: a joint-venture approach in Chongqing and Berlin (one works, one fails),[4] concession in Buenos Aires, and management contract and lease in Paris from 1860 to 2009.

The two unorthodox models analysed in the Part I provide exemplary cases of this perspective. Chongqing represents the solution the Chinese authorities have opted for to modernize the water companies. They rule out privatizations, concessions and BOT contracts, instead adopting a hybrid form: the joint-venture or semi-public company. Under this approach, the municipality – which initially owns the entire company and its assets – sells part of them to a private company; management structures, policy-setting, information and operating profits are shared. With this structure, a private partner can be introduced to run operations and contribute expertise; it also provides an answer to the classic problem of power asymmetry and represents an original contribution to the problem of regulation: its basis is not the separation between principal and agent – all information is shared, and the private partner knows that, in addition to the legal framework, there is a further framework of informal institutions. This mode of action assigns as much importance to the targets set by the municipality – equipping neighbourhoods, accelerating the modernization of the water company – as to the letter of the contract.

The Salt River Project (SRP) shares certain properties with the Chinese joint venture, although it acts in a country where the operating regime is different. We find the same departure from the standard theory, with a partnership between the public and private sectors; it operates pragmatically, providing solutions to the problems of each period. Several moments stand out from this very long history. It began, in the mid-nineteenth century, as an irrigation project backed by private reformers, with the intention of improving arid land in the western USA. At the turn of the twentieth century, the Federal government entered the picture through the promulgation of a bill, but the establishment of a second-rank institution enabled the private actors to remain involved. The land to be irrigated could be used as loan collateral in the operation, with the result that the owners of the land were involved in the management of a large infrastructure. A few years later (1905), a flood damaged the technical system and its ownership was transferred to the Federal government. In 1917, however, recurring complaints about poor local management prompted the US Reclamation Service, owner of the infrastructure, to delegate operations to the landowners' association. With no theoretical foundation, the nation of capitalism devised an original mode of operation that separates the power of ownership from management. Was this a genuine invention or the spread into the public sphere of changes that big private firms were undergoing at the time (Chandler, 1962)? In the 1920s, the initial water project was extended into the electricity sector. Dam water could be run through turbines before being used for irrigation; cities and industries were developing, and the demands were huge; by 1936 (and still today) power generation accounted for 90 per cent of revenues and made it possible to cut operating costs on the water side. After the Second World War, the company moved into a third sector – town water – in addition to irrigation and electricity production. Phoenix and Tucson were undergoing exponential urban growth, driven by the introduction of industries in the high-tech and armaments sectors. In the 2010s, however, pressure on water resources caused by increased demands (especially on groundwater) and reduced rainfall (directly affecting watercourses) prompted the SRP to become involved in large-scale water management and to initiate water conservation programmes to maintain security of supply.

One global lesson is that approaches which might seem unorthodox, as in the case of Chongqing and SRP, can produce positive results and can evolve with the context. Institutional flexibility – or the capacity of the protagonists to redefine their programme of action – is not a weakness that opens the door to all kinds of opportunists, but is instead a positive quality: the projects that survive are those that have been able to evolve. The adaptation of a model depends on the context in which the rules are implemented, and especially on the informal institutions that determine their relation to normative frames. In certain cases, respect for formal rules is considered an essential factor which, to a large degree, governs the life and legitimacy of the contract. In others, the protagonists accept more flexible institutional approaches because they know that they can rely on shared values that guide practices. The cases of Buenos Aires and Paris provide two radically opposed versions of this phenomenon. In the first case, Gouvello provides evidence of a tendency to emphasize the letter of the contract when flexibility was needed: the regulator "gradually turned into a pernickety censor", and a number of judges' rulings instituted "a trend towards the 'judicialization' of the contract". Conversely, the management contract in Paris, subsequently converted into a lease, which was not formally regulated but controlled by a "bureau" in City Hall, was flexible enough to function for a century and a half. In other words, it is difficult to prejudge the equilibrium of a system of action without understanding the relations between the various forces involved. Informal – and thus apparently fragile – ensembles may find a satisfactory balance thanks to a common ethic of public action and state expertise. It is, for example, almost impossible to explain water policies in France and China without referring to the *noblesse d'État* (Balazs, 1968; Bourdieu, 1989) shared by the two cultures. On the other hand, operations in small Indian cities demonstrate very clearly that the formalism of public policies inspired by the ideal of universalization does little to reduce the invisible barrier between a self-assured state meritocracy on the one hand, and local administrators on the other.

This diversity of possible approaches also applies to the morphology of the technical system. In the countries of the "South", solutions tend increasingly to focus on hybrid systems in which a large technical system combines with small, decentralized systems managed either by the monopoly operator or by small operators, as in Maputo, Ouagadougou, Kisumu and La Paz–El Alto. In extreme environmental contexts, a return to community-based organizations managing small, local systems may be viewed as a way to provide a balanced solution between uncontrolled urban development and scarcity of water resources.[5]

This means that the specificity of each combination should be recognized, and the analysis of the way in which it functions should include factors that may at first sight appear marginal. Indeed, researchers too often concentrate on formal institutions. As demonstrated by the small cities in Uttar Pradesh and the city of Nagpur, reforms to technical systems are not limited to large public agencies; at the end of the chain are plumbers who control access to the network. Such is the paradox: each link is important and poorly paid agents like these tradesmen, who command little respect, in fact play a key role in modernization. Similarly, although some public elites – politicians and civil servants – owe allegiance to universal principles of equality and progress, others may be guided by concepts such as apartheid and social caste systems. The methods employed to calculate tariffs may also seem neutral and objective, in that the arithmetic involved seems unquestionable. However, the major principles underpinning their calculation – marginal or average costs – are already informed by a specific view of how costs should be covered and shared among users. When calculations are based on marginal costs, consumers pay for the real cost: the volume they use, the extension to the network and connection (McCraw, 1984: 224ff.). When the average cost method is used, costs are shared among consumers connected to the network. The first method individuates (and economists believe that being aware of the costs helps consumers to avoid consuming too much), while the second introduces a form of spatial solidarity. Opting for one or other of these methods can have serious consequences. This was true in Buenos Aires, where the residents of the city's poor peripheral neighbourhoods refused to pay for their connections, which had been calculated on the basis of marginal costs. Often neglected by political decision-makers and left to professionals, the instruments of public action are, nevertheless, essential.

The conditions of collective action: competences, rules and trust

Let us start with a paradox. While the industrial policies conducted in Chongqing and Buenos Aires have been fairly similar, these operations have nevertheless met with contrasting outcomes. We find the same company, partnership-based legal structures – a joint venture in one case, a concession in the other – and the same sequence of operations on the water system. It begins with a process of mapping, complemented by field studies that are used to identify strategic nodes and to decide investment policies; in both cases, significant improvement efforts were made from the start. They were combined with maintenance operations designed to reduce water losses. Specific user-directed programmes were carried out, with a focus on billing and information services to improve day-to-day customer relations. The company's organization was modified, with the establishment of a laboratory to track the quality of the water distributed, the creation of new jobs to reflect industrial priorities and a staff training policy. The comparison could be extended to the case of Berlin. Yet, after 13 years, the Buenos Aires and Berlin partnerships stopped, whereas that in Chonqging continues. The objective differences between these cases are small. Perhaps the explanation lies less in the actual policies than in the global stance of the company, which may be described in terms of three notions: relations to politicians, relations to civil society and global positioning. Do these three sets of relations reflect a principle of trust or mistrust? And how do they develop? Over time, different dynamics can be triggered. Is the firm always and solely perceived as a commercial organization (interested in financial results) or does it ultimately become a public service enterprise capable of serving the public interest while earning a fair return? Or, to put it differently, after a certain time, having demonstrated its commitment, does the firm succeed in becoming embedded in the institutional landscape and perceived as an enterprise with the properties of an institution and not purely as a commercial organization?

Without claiming to encompass all the whys and wherefores of these complex stories, this interpretative framework nevertheless brings out certain differences. In Buenos Aires the company initially enjoyed substantial goodwill, partly originating in government, but after some five years the conditions were reversed; general policy and the economic context changed; goodwill gave way to criticism and then to mistrust. In Berlin, the trust was never there. From the start, there were conflicts between the advocates of the historical model – "strong, local, public" – and the pragmatists who wanted to open up the system. These controversies continued after the signature of the contract, and never ceased, until the final outcome: the reversion to a municipal company. In these two cases, the company never became successfully embedded in the landscape to the point of being considered something of an institution. In the Chongqing process, which began with little fanfare, the sequence was different. At the start, only a few deputy mayors backed the project, the civil servants concerned expressed misgivings but accepted the decision, while the population was not involved in the choice. The firm therefore enjoyed benevolent neutrality. It took advantage of this to apply its skills and produce results. And from this concrete foundation, its image with the population, its employees and politicians improved.

Thus, if we accept that the action revolves around the triptych of skills/rules/trust, their production and combination in the cases studied reveal several registers; these are not disembodied notions that somehow operate divorced from action. The balance between rules and trust is constantly adjusted by the reality of action. What do the protagonists do? A first sequence corresponds to a positive chain of cause and effect: operations on the network, and visible results that confirm the politicians' choice of the operator and satisfy users. A virtuous circle then develops, in which the exchanges between the authorities and the operator relate more to real things than to procedural formalism. The company can continue to develop its expertise and reduce its transaction costs: this is what has happened in Chongqing and happened in Paris between 1860 and 2009. The long history of the SRP in the USA shows us another property of this positive sequence. It is reflected in the way the company's roles and activities develop as needs change: farmland irrigation, large-scale water management with storage, electricity production from dams, distribution of town water and management of a scarce resource. The partnership lasts because it is not confined within its original purpose. Its multisectoral character, with activities in water and electricity, reinforces the operator's financial soundness and – equally importantly – helps change its status from that of a specialist organization to one with institutional properties embedded in a territory and with responsibility for essential goods. The SRP's success over a century and a half may be interpreted through this prism. By extension to this argument, it is to be wondered whether Berlin's's reversion to public status is not a reflection of the German water community's rejection of a system that separates functions – between decision-maker and operator – and places it in a position of subordination to political power. The proponents of the "strong local public model" prefer situations of co-production in a multi-utility-type industrial structure.

Controversies and what sustains them. By contrast, several of the contracts described in this volume have been characterized by the kinds of conflict that other essential services, such as telecommunications, have never had to endure. Why did these increasingly heated disputes arise in Buenos Aires, Berlin and Soweto, where practical solutions could have been found? Local political cultures do come into play. Is the action guided by a pragmatic principle such as "problem-solving" or by an ideological perspective, for example, that public goods should be managed by public companies? Classifying the cases in terms of these two principles has significant consequences for what happens next. In one case, the fact that the operator is legitimate, or more precisely is not perceived as illegitimate – in other words, enjoys a form of tacit legitimacy – is sufficient for the operator to deploy its skills and achieve results that gain the trust of elected officials and consumers (in the case of Chongqing). The principles underpinning the German "strong-local-public model" are not the same as those that characterize the pragmatism of the Chinese model. They display a political culture and an attachment to management autonomy, the downside of which is a certain mistrust of anything that is seen as differing from the local model. Controversies are an integral part of the public space, as in Soweto.

But these informal institutions need the aid of certain mediating factors to have an impact. The examples of Berlin and Soweto show clearly that opposition to private management can be fuelled by a plethora of regulations as much as second-rank institutions. For example, opponents to the "commodification" of the water service can appeal to numerous pieces of legislation in their attempt to fight the process. Indeed, opponents have used these secondary rules and rights to make their voices heard. There is no doubt that this plurality of rules at least partially explains operational difficulties in the water sector. The sector is heavily regulated by legal texts of different kinds – international agreements, national laws, local memoranda, soft law, technical standards, accounting rules and so on. These rules are subject to the principle of transparency, which means that they can be easily mobilized and are sufficiently various to feed a multiplicity of disputes. On the other hand, the decline in the service in the small towns of Uttar Pradesh has not triggered large social movements, perhaps because the problem is essentially limited to a specific professional sphere, a fact that makes it difficult to organize against inequalities in access to water. In summary, to understand why controversies escalate in some cases, in addition to the political culture, two mediating factors emerge: on the one hand, the diversity of rules and their degree of transparency; on the other, mediation (or professional management) in controversies and conflicts.

A comparison of a number of the cases also demonstrates *the importance of the time factor and of learning*. This is a particular difficulty, because by contrast with a standard market it is not enough just to manufacture a product and put it on the market; here, the market must be constructed. By running through the spectrum, we can see the complexity of a water service. When we study countries where the service dates back to the mid-nineteenth century (Berlin, Paris, SRP), others that are trying out a new form of organization (China, Nagpur), and a few that are just starting out (Africa, India), the differences between them demonstrate all the complexity of an urban water system. To supply the service to the inhabitants requires the development of a network that cannot be installed without constructing institutions; once completed, this combination constitutes a water regime. Conversely, the chapters on experiences in Africa – Maputo, Ouagadougou, Kisumu, Soweto in South Africa – and in small towns in India show the distance to go before a full service is achieved. Returning to the comparison with mobile phones, Africa was able to gear up quickly because the technologies were available and the prerequisites for market operation fairly simple – installing cellphone networks raises no particular legal problems and subscriptions quickly balance out the cost of this infrastructure. In the water sector, the process is more complex and requires time. Time to develop the whole institutional edifice, time for all the actors concerned to be comfortable with this operating regime, and for them to understand its reasons and its rules.

China, with its exploration of the different legal options (BOT, lease, management contract) between the early 1990s up to 2002, before it adopted the joint-venture solution, is a good example of handling of the time factor. Under pressure in the early 1990s from the international development institutions and the finance and consultancy industry to reform their system, the Chinese authorities declined these proposals in favour of working at their own pace. They began their reform a decade after Latin America, but this waiting time was not wasted time, since the Chinese authorities spent all those years testing, learning and investigating around the world in order to reach an opinion and an internal consensus.

The French model of urban services was clearly constructed over time. It began in the mid-nineteenth century with concessions or operating contracts, for example, in Paris and the Paris region (Defeuilley, 2014; Jacquot, 2002). The end of the nineteenth century saw the creation of a new institution which enabled towns and cities within a single urban area to share a service: this was a necessity, given the large number of units in France (Duroy, 1996; Bellanger, 2010). Before the Second World War, the principle was accepted that the sector should fund itself through its tariffs and not state subsidies. With the start of the 1950s came the transition from concessions to leases, a formula under which the main infrastructures were financed by the municipalities (or municipal groupings) and operated by private companies (Lorrain, 2008b). In the early 1980s, after these institutional adaptations, the "model" was ready to take advantage of the liberalization process favourable to the private sector. This gave the two big companies an opportunity to develop, which they did in Paris through a leasing arrangement.

The experiences around small technical systems in La Paz, in Africa and in India may also be interpreted in terms of this slow process of collective cultural adjustment. For example, the legal definition of public domain – needed to allow pipelaying – fundamentally reflects a distinction between public space and private space. This distinction is not universal, so the introduction of reforms has the effect of altering anthropological constants. In addition, the economic principle of balancing the cost of the service against a tariff paid by the consumer/user operates through a monthly or bimonthly payment. However, this demands time management in conditions of day-to-day uncertainty. The principle of the technical system introduces discipline into everyday lifestyles.

The balance between large and small technical systems

In the late 1990s, having previously supported large concessions, and without waiting for solid feedback on their results, development institutions began to take an interest in solutions based on small technical systems (Brook and Smith, 2001; Brook and Irwin, 2003; Jaglin and Zérah, 2010; Blanc and Botton, 2011). At first, these solutions seemed to solve the problems faced by large systems by involving local people, ensuring that the network served the needs of poorer neighbourhoods, and requiring less capital investment, thereby delivering lower tariffs better suited to the resources of the populations concerned. Beyond this, the question arises as to whether developing countries should attempt to reproduce the "model" that enabled industrial countries to develop water distribution networks in the late nineteenth century – a single, monopolistic network serving the entire population – or instead opt for a decentralized approach. Here again, the analysis must take on board the context in which projects are implemented.

There is no doubt that small technical systems have been successfully developed to provide the kind of service that large systems are unable to deliver. In the four small cities in Uttar Pradesh (India), in the African cities of Maputo (Mozambique), Ouagadougou (Burkina Faso) and Kisumu (Kenya), as well as in the agglomeration of La Paz–El Alto (Bolivia), there are neighbourhoods which have not been connected to the main network for a considerable length of time (see Appendix 11.1). A variety of causes may be identified: a lack of commitment on the part of the Water Agency in India; very rapid demographic growth in the African cases; a gap between white and black neighbourhoods in the case of Maputo; and geomorphological conditions in an expanding Andean city that make it difficult to construct large systems. In all of these cases, the small system seems to provide a response to a situation incompatible with the developmental prerequisites of the standard large technical system. At the same time, the data show the limits of these approaches. The cooperatives of El Alto, hand pumps in Uttar Pradesh and the small operators of Maputo are all dependent on a groundwater table which could rapidly deteriorate in quality as a result of urban development. Perhaps the answer would be to supply these systems with water from the big system, as in Ougadougou and Kisumu. The volumes of water involved are small, far lower than the international standard of 150 litres per person per day. Similarly, there is no guarantee that the commitment of the first generation of inhabitants will be mirrored by that of the second. Becoming actively involved in a water association is a constraint accepted by those who previously had nothing, but their values are not entirely consonant with those of a younger generation committed to different social relations and a different, often less localized labour market. It may be that these small systems will one day be integrated into the large system to form hybrid services, which will vary in their geometry according to the morphology and populations of the cities concerned.

Nevertheless, there is still the option of the large technical system developed by interventionist municipalities supported by the state. A comparison of performance between the different cases suggests that this is likely. The results achieved in Buenos Aires in difficult conditions and on a contract that was terminated early were substantial, even if the picture is frozen in 2001; this is true both for improvements in the technical system and in the number of people connected. If measured globally, the ratio between the populations of Buenos Aires and Maputo is 4 to 1 and between the number of connections 20 to 1, even on a cautious estimate (see Appendix 11.1). Likewise, if the mileage of pipework laid for drinking water is taken as an indicator, the comparison still favours the large system: 173 km in Maputo, 1,083 in Buenos-Aires and 668 in Chongqing. At the end of these investigations, our results converge. The authors who worked on the small technical systems describe the factors that contribute to their development; they tell us why, in certain conditions and at a certain time, it was the only solution to replace a failing public service. All, however, whether in El Alto, in Africa or in India, are lucid in questioning the durability of these experiments, and they agree on the idea that these small technical systems will survive by integrating into the existing system. The argument may be carried over to the convenience and progress brought to the inhabitants. To put things simply, it is convenient to have unlimited water on tap at home. What would it take for this to become a reality for all?

The case of Chongqing is not representative of all Chinese projects in the water sector, but it is important in two ways. The way in which joint ventures function is one explanation, but – and this is the second result – upstream factors should also be taken into account; their importance may be measured by means of term-for-term comparisons between India, Africa and various countries in Latin America. China equips its cities with large technical systems because it has a political project characterized by the desire to promote a certain type of society in which essential goods are available and accessible for everyone. Furthermore, the government has supported public policies that favour the role of public developers (Chengtou) in order to collect land rent and reinvest it in infrastructure that cannot be entirely financed by the market (public transport, tunnels, water networks, etc.). In other words, the choice between large and small technical systems is not a technical choice, and reforming contracts or public procurement will not solve the problem. In addition, the issue is linked to politics and polity – the way in which political society operates, the nature of political elites, the conception of what constitutes the difference between public and private interests, and the choice of priorities. Implementing a water distribution and sanitation network at the national level presupposes not only human and financial resources but also political will. The difficulties encountered by numerous countries poorly equipped with water and sanitation services may be viewed either as an illustration of a deficit of political commitment or as an indication that resources are being allocated to other projects and interests. There is a kind of isomorphism between politics, corruption, the absence of public interest, the allocation of collective surplus and the impossibility of governing cities, of developing large technical systems. Before looking for local solutions that demand a great deal from the populations directly concerned, this primary notion must be restated: at the top of the pyramid, driving public action forward, we find politics. Politics and polity are an integral part of the problem as well as its solution.

Appendix 11.1 How the socio-technical have been upgraded

City	*Maputo*	*Ouagadougou*	*Kisumu*	*El Alto–La Paz*	*Uttar Pradesh (4)*	*Buenos-Aires*	*Chongqing*
Total population (Mh)	2	2.7	0.5	1.5	4 × 0.02	8	1.2
Non-connected population (Mh)	1	0.20	0.32	0.24	55%, 64%, 69%, 82%	2	–
Action							
Connected population	–	32,000	15,000	–	–	1.5 M[a]	–
No. of connections	12,000	3,640	350	23,160	3,800	244,000[a]	–
Network implemented (km)	173	60+ hydrants	3	–	–	1,083 (EP)	668
Technical solution	forage	LTS[b]	LTS	Pumps and LTS	Manual pumps	LTS + STS	LTS

Notes

a Marin (2009: 46); and see Gouvello (Chapter 3, this volume). Gouvello provides data on the number of clients; the increase during the first five years is 540,000, but this figure includes the extension of the concession to cities which already had a service; for this reason we mention the more restrictive figure provided by Marin. These figures do not include people connected after 2001 under two specific programmes: MGP (100,000) and Agua + Trabajo (350,000).

b Large technical system, Small technical system.

Notes

1 This figure breaks down into 768 million people who do not have access to an "improved" source of water (referred to in international terminology as water which is protected from animal contamination); 1,300 million people who have access to water that is improved (not shared with animals) but is contaminated. Data sourced from WHO and UNICEF; for a discussion, see Payen (2013: 84).
2 See Chapter 4 (this volume) on Paris.
3 See the series of "toolkits" distributed by the PPIAF, a programme aimed at developing countries.
4 It should be observed that the institutional framework of (private) joint ventures is similar to that of French or German semi-public companies.
5 This is the case of the American West: see Chapter 2 (this volume).

References

Abbott, C. (1981) *The New Urban America: Growth and Politics in Sunbelt Cities*. Chapel Hill: The University of North Carolina Press.

Abbott, C. (1993) *The Metropolitan Frontier. Cities in the Modern America West*. Tucson and London: University of Arizona Press.

Agathe, E. (2004) Que se cache-t-il derrière les courbes de consommation d'eau? L'exemple de Paris. Communication, Latts, Ecole des Ponts, May, p. 4.

Agrikoliansky, E. (2010) Les usages protestataires du droit. In Fillieule, O., Agrikoliansky, E. and Sommier, I. (eds), *Penser les mouvements sociaux. Conflits sociaux et contestations dans les sociétés contemporaines*. Paris: La Découverte, coll. "Recherches", pp. 225–243.

Ahlers, R., Schwartz, K. and Perez Güida, M.V. (2013) The Myth of "Healthy" Competition in the Water Sector: The Case of Small-scale Water Providers. *Habitat International*, 38: 175–182.

Akhmouch, A. (2009) De l'Utilité politique de l'accusation de "pillage": le cas des multinationales de l'eau en Argentine, l'exemple de Córdoba. *Hérodote*, 3(134): 194–217.

Albo, X., Greaves, T. and Sandoval, G. (1985–1990) *Chukiyawu, la cara aymara de La Paz, t. I à IV, Nuevos Lazos con el campo*. La Paz: CIPCA.

Alcazar, L., Abdala, M.A. and Shirley, M.M. (2000) The Buenos Aires Water Concession. In Shirley, M.M. (ed.), *Thirsting for Efficiency: The Economics and Politics of Urban Water System Reform*. New York: Pergamon/The World Bank, pp. 65–102.

Annales. Histoire, Sciences Sociales (2002) no. 3, Special issue: "Politique et contrôle de l'eau dans le Moyen-Orient ancient" ["Water Policy and Control of Water in the Ancient Middle East"].

Anparthi, A. (2013) OCWL Faces 110% Liquidated Damages for Poor Water Supply. *The Times of India*, 6 June.

Arun Maira, A. (2013) India's 2% CSR Law. The First Country to Go Backwards. *Economic and Political Weekly*, 48(38): 23–25.

Asthana, A.N. (2008) Decentralisation and Corruption: Evidence from Drinking Water Sector. *Public Administration and Development*, 28(3): 181–189.

Aubriot, J. (2012) Usages militants du droit à l'eau en Afrique du Sud. Du projet Gcin Amanzi à l'affaire Mazibuko. Doctoral thesis, École nationale des ponts et chaussées.

Aubriot, J. and Moretto, L. (2013) Le droit à l'eau au Sud, un outil paradoxal pour penser le droit à la ville: entre coproduction et contestation des services urbains. In Les Cahiers d'architecture La Cambre-Horta, no. 9, *La Ville comme bien commun. Planification urbaine et droit à la ville*. Brussells: La Lettre volée/Université libre de Bruxelles.

August, J.L. (1999) *Vision in the Desert. Carl Hayden and Hydropolitics in the American Southwest.* Fort Worth: TCU Press.

Avery, C., Consoli, C., Glennon, R. and Megdal, S. (2007) Good Intentions, Unintended Consequences: The Central Arizona Groundwater Replenishment District. *Arizona Law Review*, 200(49): 339–359.

Azpiazu, D. (2010) Privatización del agua y el saneamiento en Argentina: El caso paradigmático de Aguas Argentinas S.A. *VertigO*, Hors série 7.

Babuziaux, C. and de Leyssac, C.L. (eds) (2005) *Régie, marché, partenariat, délégation. Quelle compétition pour l'amélioration du service public?* Paris: Institut de la Gestion Déléguée/La documentation Française.

Bakker, K. (2003) Archipelagos and Networks: Urbanization and Water Provision in the South. *The Geographical Journal*, 169(4): 328–341.

Bakker, K. (2010) *Privatizing Water. Governance Failure and the World's Urban Water Crisis*, Ithaca, NY, and London: Cornell University Press.

Balazs, E. (1968) *La Bureaucratie céleste. Recherches sur l'économie et la société de la Chine traditionnelle.* Paris: Gallimard.

Baldwin, R., Scott, C. and Hood, C. (eds) (1998) *A Reader on Regulation.* Oxford: Oxford University Press.

Ballard, R., Habib, A. and Valodia, I. (2006) *Voices of Protest: Social Movements in Post-Apartheid South Africa.* Pietermaritzburg: University of KwaZulu-Natal Press.

Ban, R., Gupta, M.D. and Rao, V. (2010) The Political Economy of Village Sanitation in South India: Capture or Poor Information? *The Journal of Development Studies*, 46(4): 685–700.

Barraqué, B. (ed.) (2011) *Urban Water Conflicts.* Paris: UNESCO Publishing/Taylor & Francis.

Barraqué, B. (2012) Return of Drinking Water Supply in Paris to Public Control. *Water Policy*, 14: 903–914.

Bauer, C.J. (2004) *Siren Song. Chilean Water Law as a Model for International Reform.* Washington, DC: Resources for the Future.

Beall, J., Crankshaw, O. and Parnell, S. (2000) Victims, Villains and Fixers: The Urban Environment and Johannesburg's Poor. *Journal of Southern African Studies*, 26(4): 833–855.

Beaujeu-Garnier, J., Gamblin, A. and Delobez, A. (1966) *Images économiques du monde.* Paris: Sedes.

Bellanger, E. (2010) (in collaboration with Eléonore Pineau) *Assainir l'agglomération parisienne. Histoire d'une politique publique interdépartementale de l'assainissement (XIXe-XXe siècles).* Paris: SIAAP.

Bellier Report (2002) Washington, DC: The World Bank, PPIAF.

Benjamin, S. (2005) Touts, pirates and ghosts. *Sarai Reader*, 5: 242–254.

Bercegol, R. de (2012) L'émergence des municipalités. Analyse de la réorganisation des pouvoirs issue de la décentralisation sur la gouvernance de petites villes d'Uttar Pradesh. Doctoral thesis supervised by Sylvy Jaglin (LATTS) and jointly supervised by Marie-Hélène Zérah (CSH-IRD). Paris: Université Paris-Est.

Bercegol, R. de and Desfeux, A. (2011) Alternative au service public d'eau conventionnel: l'exemple des "réseaux de groupes d'usagers" d'un bidonville de Mumbai. *Métropoles*, 9: http://metropoles.revues.org/4462.

Berenschot, W. (2010) Everyday Mediation: The Politics of Public Service Delivery in Gujarat, India. *Development and Change*, 41(5): 883–905.

Bergère, M.-C. (2013) *Chine, le nouveau capitalisme d'État.* Paris: Fayard.

Beveridge, R. (2012) Consultants, Depoliticization and Arena-shifting in the Policy Process: Privatizing Water in Berlin. *Policy Sciences*, 45(1): 57.

Bhagat, R.B. (2011) Emerging Pattern of Urbanization in India. *Economic and Political Weekly*, 46(34): 10–12.

Blair, H. (2000) Participation and Accountability at the Periphery: Democratic Local Governance in Six Countries. *World Development*, 28(1): 21–39.

Blanc, A. and Botton, S. (eds) (2011) *Water Services and the Private Sector in Developing Countries. Comparative Perceptions and Discussion Dynamics*. Paris: AFD, PPIAF.

Blanc, B., Cavé, J. and Chaponnière, E. (2009) *Les Petits Opérateurs privés de la distribution d'eau à Maputo: d'un problème à une solution?* Paris: AFD, coll. "Working Papers".

Bommier, S., Giraud, G., Renouard, C., Zérah, M.-H. and Zimmer, A. (2012) *Veolia's Project in Nagpur: Water Supply in Slums and Stakeholder Engagement*. Paris/New Delhi: ESSEC IRENÉ Institute/Centre de sciences humaines.

Bond, P. (2000) *Elite Transition. From Apartheid to Neoliberalism in South Africa*. London: Pluto Press.

Bond, P. (2004) *South Africa's Resurgent Urban Social Movements. The Case of Johannesburg, 1984, 1994, 2004*. Durban: Center for Civil Society Research Report, 22.

Bond, P. (2010) *Fighting for the Right to the City: Discursive and Political Lessons from the Right to Water*. Paper delivered at the Right to Water Conference, New York: Syracuse University, 29 March.

Botton, S. (2007) *La Multinationale et le bidonville. Privatisations et pauvreté à Buenos Aires*. Paris: Karthala.

Botton, S. and Blanc, A. (2010) *Accès de tous aux services d'eau, le rôle des petits opérateurs privés à Hô Chi Minh Ville, Vietnam*. Paris: AFD, coll. "Focales".

Bourdieu, P. (1989) *La Noblesse d'Etat. Grandes écoles et esprit de corps*. Paris: Minuit.

Bourdieu, P. and Christin, R. (1990) La construction du marché [Le champ administratif et la production de la "politique du lodgement"]. In *Actes de la recherche en sciences sociales*, Vols 81–82, March. L'économie de la maison, pp. 65–85. DOI: 10.3406/arss.1990.2927.

Boyer, R. (2002) Variété du capitalisme et théorie de la regulation. *L'Année de la régulation*, 6: 125–194.

Brook, P.J. and Irwin, T.C. (eds) (2003) *Infrastructure for Poor People. Public Policy for Private Provision*. Washington, DC: The World Bank/PPIAF.

Brook, P.J. and Smith, S.M. (eds) (2001) *Contracting for Public Services. Output-based Aid and its Applications*. Washington, DC: The World Bank/IFC.

BWB (Berliner Wasserbetriebe) *Annual Report*: 1992, 1996, 2000, 2004.

Callon, M., Lascoumes, P. and Barthe, Y. (2001) *Agir dans un monde incertain, essai sur la démocratie technique*. Paris: Seuil.

Carey, M. (2010) *In the Shadow of Melting Glaciers. Climate Change and Andean Society*. Oxford: Oxford University Press.

Carpenter, D.P. (2001) *The Forging of Bureaucratic Autonomy: Reputations, Networks, and Policy Innovation in Executive Agencies, 1862–1928*. Princeton, NJ, and Oxford: Princeton University Press.

Castro, V. (2009) *Improving Water Utility Services through Delegated Management. Lessons from the Utility and Small-Scale Providers in Kisumu, Kenya, Nairobi*. Washington, DC: WSP/The World Bank.

Cavé, J. and Blanc, A. (2011) Revue de la littérature internationale sur les POP de la distribution d'eau potable. In Blanc, A. and Botton, S. (eds), *Services d'eau et secteur privé dans les pays en développement. Perceptions croisées et dynamique des reflexions*. Paris: AFD, coll. "Recherches", pp. 327–349.

CAWP and APF (2006) Lessons from the War against Prepaid Water Meters. The Struggle against Silent Disconnections Continues… Braamfontein: Anti Privatisation Forum, August. Available at: http://apf.org.za/IMG/pdf/Final_PPM_Research_Report_-_102006-2.pdf.

Chandler, A.D. (1962) *Strategy and Structure: Chapters in the History of the American Industrial Enterprise*. Cambridge, MA: The MIT Press.

Chaponnière, E. (2008) Eau et assainissement au Mozambique. Les POP de Maputo. Des opérateurs indépendants, partenaires durables du service public de l'eau. *pS-Eau Newsletter*, 57: 6–9.

Child, J. and Yan, Y. (1999) Investment and Control in International Joint-ventures: The Case of China. *Journal of World Business*, 34(1): 3–15.

Coelho, K. (2006) Tapping In: Leaky Sovereignties and Engineered (Dis)Order in an Urban Water System. *Sarai Reader*, 6 (Turbulence): 497–509.

Coeurdray, M., Cortinas, J. and Poupeau, F. (2015) The Crossed Border Disputes over Sharing Colorado River between the Southwestern States. A Sociological Perspective on Environmental Policies. *WATERLAT Working Papers*, 2(3): 148–163.

Coing, H. (1996) Mimétisme ou métissage? *Annales des Ponts et Chaussées*, 80: 58–63.

Coing, H. (2010) Preface. In Graciela Schneier-Madanes, *L'Eau mondialisée. La gouvernance en question*. Paris: La Découverte, coll. "Recherches", pp. 11–23.

Colby, B.G. and Jacobs, K.L. (eds) (2007) *Arizona Water Policy. Management Innovations in an Urbanizing Arid Region*. Washington, DC: RFF Press.

Collignon, B. (2007) Les grands bidonvilles africains: la prochaine frontière pour les distributeurs d'eau. *Revue Politique et Parlementaire*, 1043: 79–94.

Collignon, B. and Botton, S. (2014) Les opérateurs indépendants du service de l'eau en Afrique. In Jamati, C. (ed.), *L'Afrique et l'eau*. Paris: Alpharès, pp. 117–130.

Conan, H. (2003) Scope and Scale of Small-scale Independent Private Water Providers in 8 Asian Cities. Preliminary Findings. In *Promoting Effective Water Management Policies*. Manila: ADB/RETA 6031.

Cortinas, J., Coeurdray, M. and Poupeau, F. (2015) Du Reclamation Act de 1902 aux mega-projets fédéraux. Une perspecrtive socio-historique sur la genèse des politiques hydriques aux USA: le cas de l'Ouest étasunien. *Cuadernos de Desarollo Rural*, 12(76): 135–153.

Cour des comptes (2003) *La gestion des services publics d'eau et d'assainissement*. Paris: Les éditions des Journaux officiels.

Coutard, O. (2005) Urban Space and the Development of Networks: A Discussion of the "Splintering Urbanism" Thesis. In Coutard, O., Hanley, R.E. and Zimmerman, R. (eds), *Sustaining Urban Networks. The Social Diffusion of Large Technical Systems*. London and New York: Routledge, pp. 48–64.

Coutard, O. (2010) Services urbains: la fin des grands réseaux? In Coutard, O. and Lévy, J-P. (eds), *Écologies urbaines*. Paris: Economica-Anthropos, pp. 102–129.

Cronon, W. (1991) *Nature's Metropolis: Chicago and the Great West*. New York: W.W. Norton.

Crouch, C. and Streeck, W. (eds) (1996) *Les Capitalismes en Europe*. Paris: La Découverte, coll. "Recherche".

Crozier, M. and Friedberg, E. (1977) *L'Acteur et le système*. Paris: Seuil.

Curien, N. (2000) *Economie des réseaux*. Paris: La Découverte.

Czada, R. and Lehmbruch, G. (eds) (1998) *Transformationspfade in Ostdeutschland, Beiträge zur sektoralen Vereinigungspolitik*. Frankfurt-am-Main: Campus Verlag.

Dardenne, B. (2005) *L'eau et le feu: la première Compagnie des Eaux de Paris (1777–1788)*. Paris: Editions de Venise.

Defeuilley, C. (2001) Enron. *Flux*, 46: 76–84.

Defeuilley, C. (2009) Retail Competition in Electricity Markets. *Energy Policy*, 37(2): 377–386.

Defeuilley, C. (2013) Portrait d'entreprise. La New River Company. *Flux*, 91: 56–65.

Defeuilley, C. (2014) Légitimité de l'intervention privée dans un service public. Le renouvellement du contrat des eaux de la banlieue de Paris. *Actes de la recherche en sciences sociales*, 203: 48–59.

Degrémont Beijing (1999) Documents.

Deliège, R. (1995) *Les Intouchables en Inde. Des castes d'exclus*. Paris: Imago.

Denis, É. and Marius-Gnanou, K. (2011) Toward a Better Appraisal of Urbanization in India. A Fresh Look at the Landscape of Morphological Agglomerates. *Cybergeo. European Journal of Geography* [online], Systèmes, Modélisation, Géostatistiques, doc.569, uploaded on 28 November 2011 (http://cybergeo.revues.org/24798; DOI: 10.4000/cybergeo.24798).

Dezalay, Y. and Garth, B., (1998) Le "*Washington consensus*". Contribution à une sociologie de l'hégémonie du néolibéralisme. *Actes de la recherche en sciences sociales*, 121–122: 3–22.

Dhaliwal, S. (2004) *Urban Infrastructure Development in Small and Medium Towns*. New Delhi: Deep and Deep Publications.

Duany, A., Plater-Zyberk, E. and Speck, J. (2001) *Suburban Nation: The Rise of Sprawl and the Decline of the American Dream*. New York: Macmillan.

Dupont, V. (2002) Le monde des villes. In Saglio-Yatzimirsky, M.-C. (ed.), *Population et développement en Inde*. Paris: Ellipses, pp. 55–84.

Dureau, F., Gouëset, V. and Mesclier, E. (eds) (2006) *Géographies de l'Amérique latine*. Rennes: Presses Universitaires de Pennes.

Duroy, S. (1996) *La distribution d'eau potable en France*. Paris: LGDJ.

Eales, K. and Smith, L. (2011) Water and sanitation. In *Accountability in Public Services in South Africa*. Washington, DC: The World Bank, March, pp. 87–103.

Easter, K.W., Feder, G., Le Moigne, G. and Duda, A.M. (1993) *Water Resources Management*. A World Bank Policy Paper. Washington, DC: The World Bank.

Eichholz, M. (2010) *Regulation und Krise der Wasserversorgung in La Paz (Bolivien). Analyse des Potentials dezentraler Wasserkooperativen aus sozial-ökologischer Perspektive*. Bonn: Diplomarbeit, Geographisches Institut der Rheinischen Friedrich-Wilhelms-Universität, IRD.

Euzen, A. (2004) Que se cache-t-il derrière les courbes de consommation d'eau? L'exemple de Paris. 15èmes Journées scientifiques de l'Environnement – Usages de l'eau: sygergies et conflits, Créteil.

Faudry, D. (1995) La concession des services d'eau potable et l'assainissement de Buenos Aires. In Lorrain, D. (ed.), *Gestions Urbaines de l'Eau*. Paris: Economica, pp. 117–133.

Faudry, D. (1997) The Concession of Buenos Aires' Water Supply and Waste Water Services (The Largest Water Services Concession in the World). In Lorrain, D. (ed.), *Urban Water Management*. Levallois-Perret: Hydrocom, pp. 129–147.

Franceys, R. and Gerlach, E. (eds) (2008) *Regulating Water and Sanitation for the Poor*

(Economic Regulation for Public and Private Partnerships). London: Sterling, VA, Earthscan.

Franco, J., Mehta, L. and Veldwisch, G.J. (2013) The Global Politics of Water Grabbing. *Third World Quarterly*, 34(9). Available at: www.ids.ac.uk/publication/the-global-politics-of-water-grabbing.

Fudala, J. (2001) *Historic Scottsdale, A Life from the Land*. San Antonio, TX: Historical Publishing Network.

Gaggero, J., Gerchunoff, P., Porto, A. and Urbiztondo, S. (1992) Algunas consideraciones sobre la privatización de Obras Sanitarias de la Nación. *Estudios*, 63.

Gammage, G. (1999) *Phoenix in Perspective. Reflections on Developing the Desert*. Tempe: Herberger Center for Design Excellence.

Garnier, B. *et al.* (1966) *Images économiques du monde*. Paris: Sedes.

Garreau, J. (1991) *Edge Cities: Life on the New Frontier*. New York: Doubleday.

Gentle, L. (2008) Afrique du Sud: "apartheid social" et contestations populaires. In *État des résistances dans le Sud – 2008*. Louvain-la-Neuve/Paris: Centre tricontinental/ Syllepse, coll. "Alternatives sud". Available at: http://cetri.domainepublic.net/ Afrique-du-Sud-apartheid-social-et.

Ginisty, K. (2009) *Les petits opérateurs privés d'approvisionnement en eau à Maputo. Inscription spatiale et acteurs du territoire*. Internal report. Paris: AFD.

Gober, P. (2006) *Metropolitan Phoenix: Place Making and Community Building in the Desert*. Philadelphia: University of Pennsylvania Press.

Gottlieb, R. (1988) *A Life of its Own: The Politics and Power of Water*. San Diego: Harcourt Brace Jovanovich.

Gottlieb, R. and Fitzsimmons, M. (1991) *Thirst for Growth. Water Agencies as Hidden Government in California*. Tucson: University of Arizona Press.

Goubert, J.-P. (1986) *La Conquête de l'eau, (L'avènement de la santé à l'âge industriel)*. Paris: Robert Laffont.

Gouvello, B. de (2001) *Les services d'eau et d'assainissement en Argentine à l'heure néolibérale. La mondialisation des "modèles" à l'épreuve du territoire*. Paris: L'Harmattan.

Gouvello, B. de and Fournier, J.-M. (2002) Résistances locales aux "privatisations" des services de l'eau: les cas de Tucuman (Argentine) et Cochabamba (Bolivie). *Autrepart*, 21(1): 69–82.

Graham, S. and Marvin, S. (2001) *Splintering Urbanism. Networked Infrastructures, Technological Mobilities and the Urban Condition*. London and New York: Routledge.

Guasch, J.L. (2004) *Granting and Renegotiating Infrastructure Concessions. Do It Right*. Washington, DC: World Bank Institute Studies.

Gupta, S., Kumar, S. and Sarangi, G. (2011) Does Greater Autonomy Improve Performance? Evidence from Water Service Providers in Indian Cities. Centre for Development Economics, Department of Economics, Delhi School of Economics, *Working Paper no. 205*.

Guttman, D. and Willner, B. (1976) *The Shadow Government*. New York: Pantheon Books.

Hailu, D., Rendtorff-Smith, S. and Tsukada, R. (2011) *Small-scale Water Providers in Kenya: Pioneers or Predators?* New York: UNDP.

Hall, P.A. and Soskice, D. (2002) Les variétés du capitalisme. *L'Année de la régulation*, 6: 47–124.

Hardin, G. (1968) The Tragedy of the Commons. *Science*, 13 December: 1243–1248.

Hardy, S. (2009) La vulnérabilité de l'approvisionnement en eau dans l'agglomération pacénienne: le cas du sous-système El Alto. *Cybergeo: European Journal of Geography*, Dossiers, Vulnérabilités urbaines au sud, 457, put online on 20 May (www.cybergeo.eu/index22270.html).

Hardy, S. (2013) *Atlas de la vulnérabilité de l'agglomération de La Paz*. Marseille: Editions de l'IRD.

Hardy, S. and Poupeau, F. (2014) L'auto-organisation de la gestion urbaine de l'eau. La fonction des coopératives dans le grand système La Paz/El Alto. *Actes de la recherche en sciences sociales*, 203: 86–105.

Harris, C. (2003) *Private Participation in Infrastructure in Developing Countries, Trends, Impacts, and Policy Lessons*. Washington, DC: World Bank Working Paper no. 5.

Harris, C., Hodges, J., Schur, M. and Shukla, P. (2003) *Infrastructure Projects – A Review of Canceled Private Projects. Public Policy for the Private Sector*. Washington, DC: The World Bank.

Hayes, S.P. (1999) *Conservation and the Gospel of Efficiency*. Pittsburgh, PA: University of Pittsburgh Press.

Hibou, B. (1998) Retrait ou redéploiement de l'Etat? *Critique internationale*, 1: 151–168.

Himanshu, (2006) Urban Poverty in India By Size Class of Towns: Level, Trends and Characteristics. Paper presented at the seminar "The Multi Dimensions of Urban Poverty in India". Mumbai: Indira Gandhi Institute of Development Research, 6–7 October.

Hofstadter, R. (1955) *The Age of Reform*. New York: Vintage Books.

Huchon, A. and Tricot, G. (2008) *Between Citizens and Institutions: The Dynamics of the Integration of Water Supply and Sanitation Services in Hyderabad*. New Delhi: The French Research Institutes in India, CSH Occasional Paper, 22.

Hüesker, F. (2011) *Kommunale Daseinsvorsorge in der Wasserwirtschaft. Auswirkungen der Privatisierung am Beispiel der Wasserbetriebe des Landes Berlin*. Munich: Oekom Verlag.

Hughes, T.P. (1983) *Networks of Power. Electrification in Western Society, 1880–1930*. Baltimore, MD: The Johns Hopkins University Press.

Hundley, N. (1975) *Water and the West. The Colorado River Compact and the Politics of Water in the American West*. Berkeley, Los Angeles, and London: University of California Press.

Hundley, N. (2001) *The Great Thirst. Californians and Water. A History*. Berkeley, Los Angeles, and London: University of California Press.

Hydroconseil (2013) *Évaluation du projet "quartiers périphériques"*. Avignon.

Ingold, A. (2008) Les sociétés d'irrigation: bien commun et action collective. *Entreprises et histoire*, 50: 19–35.

Iribarne, P. d' (2009) *L'Épreuve des différences. L'expérience d'une entreprise mondiale*. Paris: Seuil.

Israel, A. (1993) *La Gestion de l'infrastructure, les enjeux des années 90*. Document de synthèse de la Banque mondiale, 171F.

Jacquot, A. (2002) La Compagnie générale des eaux 1852–1952: un siècle, des débuts à la renaissance. *Entreprises et Histoire*, 30: 32–44.

Jaglin, S. (2003a) *Réseaux et fragmentation urbaine: services d'eau en Afrique subsaharienne*, HDR. Paris: Université Paris I-Panthéon-Sorbonne.

Jaglin, S. (2003b) Service d'eau et construction métropolitaine au Cap (Afrique du Sud): les difficultés de l'intégration urbaine. *Revue française d'administration publique*, 107: 433–445.

Jaglin, S. (2005) *Services d'eau en Afrique subsaharienne. La fragmentation urbaine en question*. Paris: CNRS Éd.

Jaglin, S. and Zérah, M.-H. (2010) Eau des villes: repenser des services en mutation. Introduction. *Revue Tiers Monde*, 203(3): 7–22.

Jaglin, S., Clément Répussard, R. and Belbeoc'h A. (2011) Decentralisation and Governance of Drinking Water Services in Small West African Towns and Villages (Benin, Mali, Senegal): The Arduous Process of Building Local Governments. *Canadian Journal of Development Studies*, 32(2): 119–138.

Jensen, M.C. and Meckling, H.W. (1976) Theory of the Firm: Managerial Behavior, Agency Costs and Ownership Structure. *Journal of Financial Economics*, 3(4): 305–360.

Johnson, R. (1977) *Central Arizona Project (1918–1968)*. Tucson: University of Arizona Press.

Jullien, F. (1998) *Un sage est sans idée, ou l'autre de la philosophie*. Paris: Seuil.

Kalamkar, S.S. and Shroff, S. (2011) Impact of Rehabilitation Package in Suicide-prone Districts of Vidarbha. *Economic and Political Weekly*, 5(46): 10–13.

Keane, J. (1991) Managing Water Supply Variability: The Salt River Project. In *Managing Water Resources in the West Under Conditions of Climate Uncertainty: Proceedings*, pp. 303–323.

Kernen, A. (1999) Shenyang: l'avant-garde du socialisme à l'heure des privatisations. In Hibou, B. (ed.), *La Privatisation des États*. Paris: Karthala, pp. 109–129.

Kessides, C. (1993) *Institutional Options for the Provision of Infrastructure*. Washington, DC: World Bank Discussion Papers, 212.

Kessides, I. (2004) *Reforming Infrastructure. Privatization, Regulation, and Competition*. Washington, DC and London: The World Bank/Oxford University Press.

Klein, M. and Roger, N. (1994) *Back to the Future. The Potential in Infrastructure Privatization*. The Amex Bank Review Prize Essays. Oxford: Oxford University Press.

Kocka, J. (ed.) (1996), *Les Bourgeoisies européennes au XIXe siècle*. Paris: Belin.

Kuisel, R.F. (1984) *Le capitalisme et l'Etat en France; modernisation et dirigisme au 20e siècle*. Paris: Gallimard, bibliothèque des histoires.

Kumpel, E. and Nelson, K.L. (2013) Comparing Microbial Water Quality in an Intermittent and Continuous Piped Water Supply. *Water Research*, 47(14): 5176–5188.

Kundu, A. (2011) Method in Madness: Urban Data from 2011 Census. *Economic and Political Weekly*, 46(40): 13–16.

Kundu, A. and Thakur, S. (2006) Access to Drinking Water in Urban India: An Analysis of Emerging Spatial Pattern in the Context of New System of Governance. In Reddy, R. and Dev, M. (eds), *Managing Water Resources: Policies, Institutions and Technologies*. Oxford: Oxford University Press.

Kundu, D. (2009) Elite Capture and Marginalization of the Poor in Participatory Urban Governance: A Case of Resident Welfare Association in Metro Cities. In Ministry of Housing and Urban Poverty Alleviation, Government of India, *India: Urban Poverty Report 2009*. New Delhi: Oxford University Press, pp. 271–286.

Kundu, D. and Dibyendu, S. (2011) Redefining the Inclusive Urban Agenda in India. *Economic and Political Weekly*, 46(5): 55–63.

Kupel, D. (2003) *Fuel for Growth. Water and Arizona's Urban Environment*. Tucson: University of Arizona Press.

Laffont, J-J. and Martimort, D. (2002) *The Theory of Incentives. The Principal–Agent Model*. Princeton, NJ: Princeton University Press.

Lane, J. (2012) Barriers and Opportunities for Sanitation and Water for All, as Envisaged by the New Delhi Statement. In Nicol, A., Mehta, L. and Allouche, J. (eds), *IDS Bulletin 43.2* ('Some for All'? Politics and Pathways in Water and Sanitation). Available at: http://onlinelibrary.wiley.com/doi/10.1111/j.1759-5436.2012.00302.x/pdf.

Lascoumes, P. and Le Galès, P. (2004) L'action publique saisie par ses instruments. In Lascoumes, P. and Le Galès, P. (eds), *Gouverner par les instruments*. Paris: Presses de Sciences Po., pp. 11–44.

Lascoumes, P. and Simard, L. (2011) L'action publique au prisme de ses instruments. *Revue française de science politique*, 61(1): 5–22.

Le Goff, J. (ed.) (1989) *l'Etat et les pouvoirs*. In Burguière, A. and Revel, J. (eds), *Histoire de la France*. Paris: Seuil.

Lee, K.S. and Anas, A. (1989), *Manufacturers' Responses to Infrastructure Deficiencies in Nigeria: Private Alternatives and Policy Options*. Policy Research Working Paper WPS 325, The World Bank.

Lequin, Y. (1983) Les citadins et leur vie quotidienne. In Duby, G. (ed.), *Histoire de la France urbaine*, t. 4, *La ville de l'âge industriel: le cycle haussmannien*. Paris: Seuil, pp. 275–355.

Lévy-Leboyer, M. and Casanova, J.C. (eds) (1991) *Entre l'Etat et le marché (L'économie française des années 1880 à nos jours)*. Paris: Gallimard, nrf.

Llorente, M. and Zérah, M.-H. (2005) Enjeux d'eau en Inde. Des Effets allocatifs et redistributifs complexes. *Sciences de la société*, no. 64. Toulouse: Presses universitaires du Mirail, pp. 175–192.

Lobina, E. and Hall, D. (2000) Public Sector Alternatives To Water Supply And Sewerage Privatisation: Case Studies. *International Journal of Water Resources Development*, 16(1): 35–55.

Logan, M.F. (2006) *Desert Cities: The Environmental History of Phoenix and Tucson*. Pittsburg: University of Pittsburgh Press.

Lorrain, D. (1992) The French Model of Urban Services. *West European Politics*, 15(2): 77–92.

Lorrain, D. (2004) Les pilotes invisibles de l'action publique, le désarroi du politique? In Lascoumes, P. and Le Galès, P. (eds), *Gouverner par les instruments*. Paris: Presses de Sciences Po., pp. 163–197.

Lorrain, D. (2005) Urban Capitalisms: European Models in Competition. *International Journal of Urban and Regional Research*, 29(2): 231–267.

Lorrain, D. (2007) The Local–Global Firm: Lyonnaise des Eaux, 1980–2004. *Sociologie du travail*, 49(5): 90–109.

Lorrain, D. (2008a) Les institutions de second rang. *Entreprises et histoire*, 50: 6–13.

Lorrain, D. (2008b) La naissance de l'affermage: coopérer pour exister. *Entreprises et histoire*, 50: 67–85.

Lorrain, D. (2009) Compétitions dans le secteur de l'eau, hal-00995122.

Lorrain, D. (2010) Portrait d'entreprise. Blue red chips: les compagnies d'eau en Chine. *Flux*, 82: 64–80.

Luo, Y. (2009) Are We on the Same Page? Justice Agreement in International Joint-ventures. *Journal of World Business*, 44(4): 383–396.

Lyonnaise des Eaux (2008) *12 Undertakings for Sustainable Water Management*. Assessment Report.

McCraw, T.K. (1984) *Prophets of Regulation: Charles Francis Adams, Louis D. Brandeis, James M. Landis, Alfred E. Kahn*. Cambridge, MA: Belknap Press/Harvard University Press.

McDonald, D.A. and Pape, J. (2002) *Cost Recovery and the Crisis of Service Delivery in South Africa*. London and New York: HSRC Publishers/Zed Books.

McGranahan, G. and Satterthwaite, D. (2006) *Governance and Getting the Private Sector to Provide Better Water and Sanitation Services to the Urban Poor*. London: IIED.

McKenzie, D. and Ray, I. (2009) Urban Water Supply in India: Status, Reform Options and Possible Lessons. *Water Policy*, 11(4): 442–460.

McKinley, D.T. (2004) *Democracy and Social Movements in South Africa*. Paper delivered at the UKZN School of Development Studies Conference: "Reviewing the First Decade of Development and Democracy in South Africa". Durban, 22–24 October.

Mahadevia, D. and Mukherjee, A. (2003) *Municipal Finances and Development in Small and Medium Towns: Comparing Gujarat, Punjab and Andhra Pradesh*. Working Paper Series, 18. New Delhi: Institute For Human Development.

Mairie de Paris (2012) Annual reports on the price and quality of the public potable and wastewater services.

Mann, D.E. (1963) *The Politics of Water in Arizona*. Tucson: University of Arizona Press.

Marin, P. (2009) *Public–Private Partnerships for Urban Water Utilities. A Review of Experiences in Developing Countries*. Trends and Policy Options, 8. Washington, DC: The World Bank/PPIAF.

Marin, P. and Izaguirre, A.K. (2006) Private Participation in Water: Toward a New Generation of Projects? *Gridlines*, 14. Washington, DC: The World Bank/PPIAF.

Marin, P., Mas, J.-P. and Palmer, I. (2009) Using a Private Operator to Establish a Corporatized Public Water Utility. The Management Contract for Johannesburg Water. World Bank, Water Working Notes, 20, Washington, DC.

Mathew, G. (ed.) (2000) *Status of Panchayati Raj in the States and Union Territories of India*. New Delhi: Institute of Social Sciences (ISS)/Concept Publishing Company.

Matsinhe, N.P., Juízo, D., Macheve, B. and dos Santos, C. (2008a) Regulation of Formal and Informal Water Service Providers in Peri-urban Areas of Maputo, Mozambique. *Physics and Chemistry of the Earth*, 33(8–13): 841–849.

Matsinhe, N.P., Juízo, D., Rietvel, L.C. and Persson, K.M. (2008b) Water Services with Independent Providers in Peri-urban Maputo: Challenges and Opportunities for Long-term Development. *Water SA*, 34(3): 411–420.

Mayhew, B. (2008) Accounting Research on Co-operatives; Discussion Paper Prepared for The University of Wisconsin Center for Cooperatives as part of their Discussion Paper Grant.

Mayntz, R. and Hughes, T.P. (eds) (1988) *The Development of Large Technical Systems*. Frankfurt-am-Main and Boulder, CO: Campus Verlag and Westview Press.

Mehta, S. (2004) *Maximum City: Bombay Lost and Found*. New York: Alfred A. Knopf.

Menon, A. (2012) The Rise and Rise of Nitin Gadkari. *India Today*, 28 October.

Merklen, D. (2008) L'eau et le travail. Le marché de l'eau potable, entre politiques sociales et traditions politiques, Argentine 2004–2006. *Entreprises et Histoire*, 50(April): 52–66.

Meschi, P.-X. and Riccio, E.L. (2008) Country Risk, National Cultural Differences Between Partners and Survival of International Joint-ventures in Brazil. *International Business Review*, 17(3): 250–266.

Mestrallet, G. (2001) La vraie bataille de l'eau. Open letter in *Le Monde*, 26 October.

Meublat, G. (2001) La rénovation des politiques de l'eau dans les pays du Sud. *Revue Tiers Monde*, 166: 249–257.

Minujin, A. and Lopez, N. (1995) Au sujet des pauvres et des vulnérables: le cas argentin. *Tiers-Monde*, 36(142): 365–381.

Mishra, J.S. (2002) *A Quest for Dream Cities*. New Delhi: Har-Anand Publications.

Misra, K. (2014) From Formal-informal to Emergent Formalisation: Fluidities in the Production of Urban Waterscapes. *Water Alternatives*, 7(1): 15–34.

Mohl, R.A. (1990) *Searching for the Sunbelt: Historical Perspectives on a Region*. Knoxville: University of Tennessee Press.

Molotch, H. and Logan, J. (1987) *Urban Fortunes: The Political Economy of Place*. Berkeley: University of California Press.

Mouchard, D. (2002) Les mobilisations des "sans" dans la France contemporaine: l'émergence d'un "radicalisme autolimité"? *Revue française de science politique*, 52(4): 425–447.

Mouchard, D. (2003) Une ressource ambivalente: les usages du répertoire juridique par les mouvements de "sans". *Mouvements*, 29: 55–59.

Mount, J., Hanak, E., Lund, J., Frank, R., Gartrel, G., Gray, B., Moyle, P. and Thompson, B. (2015) *California's Water: Managing Drought*. San Francisco: Public Policy Institute of California. PPIC Water Policy Center.

Mubangizi, J.C. and Mubangizi, B.C. (2005) Poverty, Human Rights Law and Socio-economic Realities in South Africa. *Development Southern Africa*, 22(2): 277–290.

Natale, A. (1993) *Privatizaciones en privado*. Buenos Aires: Planeta.

Needham, A. (2014) *Power Lines: Phoenix and the Making of the Modern Southwest*. Princeton, NJ: Princeton University Press.

Nigam, J. (1990) *Kushinagar Development Project under Urban Water Supply Reorganization Scheme*. Government of Uttar Pradesh.

Nigam, J. (2005) *Chandauli Nagar Panchayat Water Supply Reorganization Scheme under UIDSSMT*. Government of Uttar Pradesh.

Nora, P. (2013) *Recherches de la France*. Paris: Gallimard nrf.

North, D.C. (1990) *Institutions, Institutional Change and Economic Performance*. Cambridge: Cambridge University Press.

OECD (2011) OECD *Guidelines for Multinational Enterprises*. Paris: OECD (http://dx.doi.org/10.1787/9789264115415-en).

Oelmann, M., Böschen, I., Kschonz, C. and Gernot Müller, G. (2011) *10 Jahre Wasserpartner Berlin. Eine Bilanz der öffentlich-privaten Partnerschaft zwischen dem Land Berlin, RWE Aqua und Veolia Wasser*. Bad Honnef: WIK-Consult, 2009.

Ordoqui Urcelay, B. (2007) *Servicios de agua potable y alcantarillado en la ciudad de Buenos Aires, Argentina: factores determinantes de la sustentabilidad y el desempeño*. Santiago de Chile: CEPAL, Natural Resources and Infrastructure Series no. 126.

Ostrom, E. (1990) *Governing the Commons. The Evolution of Institutions for Collective Action*. New York: Cambridge University Press.

Ostrom, E. (2009) A General Framework for Analyzing Sustainability of Social-ecological Systems. *Science*, 325(24): 419–423.

Pairault, T. (2008) *Petite introduction à l'économie de la Chine*. Paris: Éditions des archives contemporaines/Agence universitaire de la francophonie.

Parnell, S. (2005) Constructing a Developmental Nation – The Challenge of Including the Poor in the Post-apartheid City. *Transformation. Critical Perspectives on Southern Africa*, 58: 20–44.

Parnell, S. and Boulle, J. (2008) Utopie urbaine et gouvernement local. La stratégie de développement économique et humain au Cap. In Dubresson, A. and Jaglin, S.

(eds), *Le Cap après l'apartheid. Gouvernance métropolitaine et changement urbain*. Paris: Karthala, pp. 251–273.

Parnell, S. and Pieterse, E. (2010) The "Right to the City": Institutional Imperatives of a Developmental State. *International Journal of Urban and Regional Research*, 34(1): 146–162.

Payen, G. (2013) *De l'eau pour tous. Abandonner les idées reçues, affronter les réalités*. Paris: Armand Colin.

Péchu, C. (1996) Quand les "exclus" passent à l'action. La mobilisation des mal-logés. *Politix*, 34: 114–133.

Pedregal, B., Hernández-Mora, N., Cabello, V., Limones, N. and Del Moral, L. (2015) Information and Knowledge for Water Governance in the Networked Society. *Water Alternatives*, 8(2): 1–19.

Pierbattisti, D. (2006) *La privatización de ENTel y la transformación de las identidades en el trabajo: génesis del dispositivo neoliberal en Argentina. 1990–2001*. Buenos Aires: Instituto Gino Germani, Documentos de Jóvenes Investigadores, no. 10.

Pincetl, S. (2011) Urban Water Conflicts in the Western United States. In Barraque, B. (ed.), *Urban Water Conflicts*. Paris: UNESCO, pp. 237–246.

Pisani, D.J. (1982) State vs. Nation: Federal Reclamation and Water Rights in the Progressive Era. *Pacific Historical Review*, 51(3): 265–282.

Pisani, D.J. (1984) *From the Family Farm to Agribusiness: The Irrigation Crusade in California and the West, 1850–1931*. Berkeley: University of California Press.

Pisani, D.J. (1994) George Maxwell, the Railroads, and American Land Policy, 1899–1904. *Pacific Historical Review*, 63(2): 177–202.

Pisani, D.J. (2002) *Water and American Government: The Reclamation Bureau, National Water Policy, and the West, 1902–1935*. Berkeley and London: University of California Press.

Poupeau, F. (2008) *Carnets boliviens, 1999–2007. Un goût de poussière*. Paris: Aux lieux d'être.

Poupeau, F. (2010) De la migración rural à la movilidad intraurbana. Una perspectiva sociológica sobre las desigualdades de acceso al agua en El Alto (Bolivia). In Poupeau, F. and González-Andricaín, C. (eds), *Modelos de gestión del agua en los Andes*. Lima: IFEA/PIEB, pp. 243–280.

Poupeau, F. (2011a) L'eau de la *Pachamama*. Commentaires sur l'idée d'indigénisation de la modernité. *L'Homme*, 198–199: 247–276.

Poupeau, F. (2011b) L'envers de la "ville aymara". Migration rurale, mobilité intraurbaine et mobilisations politiques à El Alto (Bolivie). *Revue d'études en agriculture et en environnement*, 92(4): 417–440.

Reisner, M. (1985) *Cadillac Desert: The American West and its Disappearing Water*. New York: Penguin Books.

Rice, B.R. and Bernard, R.M. (1984) *Sunbelt Cities: Politics and Growth Since World War II*. Austin: University of Texas Press.

Rosanvallon, P. (1989) Etat et société (du XIXe siècle à nos jours), dans Le Goff, J. (dir.), *l'Etat et les pouvoirs*. In Burguière, A. and Revel, J. (eds), *Histoire de la France*. Paris: Seuil, pp. 491–617.

Rosanvallon, P. (2004) *Le modèle politique français, La société politique contre le jacobinisme de 1789 à nos jours*. Paris: Seuil.

Ross, A. (2011) *Bird on Fire: Lessons from the World's Least Sustainable City*. New York: Oxford University Press.

Ruet, J. (2002) Mythologie indienne. *Courrier de la planète*, 70: 19–21.

Ruet, J., Saravanan, V.S. and Zérah, M.-H. (2002) *The Water and Sanitation Scenario in Indian Metropolitan Cities: Resources and Management in Delhi, Calcutta, Chennai, Mumbai.* CSH Occasional Paper, The French Research Institutes in India, 6.

Ruiz-Meyer, F. and van Ginneken, F. (2006) *Consumer Cooperatives: An Alternative Institutional Model for Delivery of Water Supply and Sanitation Service?* Working Notes no. 5, Infrastructure Network of The World Bank.

Sabatier, P. and Jenkins, H. (1993) *Policy Change and Learning: An Advocacy Coalition Approach.* Boulder, CO: Westview Press.

Saleth, R.M. (1996) *Water Institutions in India: Economics, Law and Policy.* New Delhi: Commonwealth Publishers.

Sanjuan, T. (ed.) (2006) *Dictionnaire de la Chine contemporaine.* Paris: Armand Colin.

Satterthwaite, D. (2006) Outside the Large Cities, the Demographic Importance of Small Urban Centres and Large Villages in Africa, Asia and Latin America. Human Settlements Discussion Paper, London: IIED.

Scarborough, V.L., Valdes Jr., F. and Dunning, N. (eds) (2003) *Heterarchy, Political Economy and the Ancient Maya. The Three Rivers Region of the East-Central Yucatán Peninsula.* Tucson: University of Arizona Press.

Schipper, J. (2008) *Disappearing Desert: The Growth of Phoenix and the Culture of Sprawl.* Norman: University of Oklahoma Press.

Schulman, B.J. (1991) *From Cotton Belt to Sunbelt: Federal Policy, Economic Development, and the Transformation of the South, 1938–1980.* New York: Oxford University Press.

Schulte, E.B. (1999a) Wasserbetriebe: Konsortien mit guten Chancen. *Berliner Zeitung*, 13 February: 35.

Schulte, E.B. (1999b) Wasserbetriebe: Vivendi besteht auf Ausgleichgarantie des Landes. *Berliner Zeitung*, 29 October: 32.

Schwalbach, J., Schwerk, A.J. and Smuda, D. (2011) *Berwertung der Rekommunalisierung der Berliner Wasserbetriebe.* Berlin: Humboldt-Universität zu Berlin, Institut für Management, 28 March.

Scott, C.A. and Pasqualetti, M.J. (2010) Energy and Water Resources Scarcity: Critical Infrastructure for Growth and Economic Development in Arizona and Sonora. *Natural Resources Journal*, 50: 645–682.

Seekings, J. (2007) Poverty and Inequality after Apartheid. CSSR Working Paper 200. Rodebosch: Centre for Social Science Research, University of Cape Town.

Serageldin, I. (1994) Promoting Sustainable Development: Toward a New Paradigm. In Serageldin, I. and Steer, A. (eds), *Valuing the Environment: Proceedings of the First Annual International Conference on Environmentally Sustainable Development.* ESD, Series 2. Washington, DC: The World Bank, pp. 13–21.

Shadegg, S. (1969) *Century One: 1869–1969. One Hundred Years of Water Development in the Salt River Valley.* Phoenix: Private Editor.

Shastri, P. (2011) *How India's Small Towns Live (or Die). Making Sense of Municipal Finances.* New Delhi: Academic Foundation/PRIA.

Sheridan, T.E. (2012) *Arizona: A History.* Tucson: University of Arizona Press.

Shermer, E.T. (2013) *Sunbelt Capitalism: Phoenix and the Transformation of American Politics.* Philadelphia: University of Pennsylvania Press.

Shirley, M. (ed.) (1995) *Bureaucrats in Business. The Economics and Politics of Government Ownership.* A World Bank Policy Research Report. New York: Oxford University Press.

Si, S.X. and Bruton, G. (2003) Beyond National Boundaries: Considering the View from Both Sides of the Table. *Organizational Dynamics*, 32(4): 384–395.

Smith, C. (1972) *Salt River Project*. Tucson: University of Arizona Press.

Smith, L. (2004) The Murky Waters of the Second Wave of Neoliberalism: Corporatization as a Service Delivery Model in Cape Town. *Geoforum*, 35(3): 375–393.

Smith, L. (2006) Neither Public nor Private. Unpacking the Johannesburg Water Corporatization Model. Social Policy and Development Programme Paper, 27, UNRISD.

Snell, S. (1998) *Water and Sanitation Services for the Urban Poor. Small-Scale Providers: Typology and Profiles*. Washington, DC: The World Bank, World Bank Policy, Research Working Paper.

Solo, T.M. (2003) Proveedores independientes de agua potable en América Latina: El otro sector privado en los servicios de abastecimiento de agua. Washington, DC: World Bank Policy, Research Working Paper.

Solomon, R.C. and Hanson, K.R. (1985) *It's Good Business*. New York: Atheneum.

SRP (2014a) *Making Connections*. SRP Annual Report.

SRP (2014b) *Combined Financial Statements. Report of Independant Auditors (April 30)*. SRP Annual Report.

SRP (2014c) *Waterways, News about SRP Water Stewardship*. Winter.

Stiglitz, J.E. (2001) *Information and the Change in the Paradigm in Economics*. Stockholm: The Nobel Foundation.

Stiglitz, J.E. (2002) *Globalization and Its Discontents*. New York: W.W. Norton.

Stiglitz, J.E.W. (2003) *The Roaring Nineties. Why We're Paying the Price for the Greediest Decade in History*. New York: W.W. Norton.

Stoffaës, C. (1995) *Services publics. Question d'avenir*. Paris: Editions Odile Jacob, La Documentation française.

Suez-Lyonnaise des Eaux (2008) *12 Undertakings for Sustainable Water Management*. Assessment Report.

Suez Environnement (2010a) *Ethics Charter*. Paris: La Défense.

Suez Environnement (2010b) *Engagements and Performances. Développement durable*. Paris: La Défense. Available at: www.suez-environnement.fr/fr/developpement-durable/performances/rapports-engagements-et-performances/.

Summit, A.R. (2013) *Contested Waters. An Environmental History of the Colorado River*. Boulder: University Press of Colorado.

Supreme Audit Institution of India (2013) *Performance Audit on Jawaharlal Nehru National Urban Renewal Mission, Report No. 15 of 2012–13 for the Period Ended March 2012*. New Delhi: Ministry of Urban Development/Ministry of Housing and Urban Poverty Alleviation, Comptroller and Auditor General of India.

Svampa, M. (1999) What Remains of Peronist Culture in Late 1990s Argentina? *Problèmes de l'Amérique Latine*, Paris, October/December: 2540.

Swyngedouw, E. (2004) *Social Power and Urbanization of Water. Flows of Power*. Oxford: Oxford University Press.

Tarr, J.A. (1989) Infrastructure and City-building in the Nineteenth and Twentieth Centuries. In Hays, S.P. (ed.), *City at the Point. Essays on the Social History of Pittsburgh*. Pittsburgh: University of Pittsburgh Press.

Tarr, J. and Dupuy, G. (eds) (1988) *Technology and the Rise of the Networked City in Europe and America*. Philadelphia, PA: Temple University Press.

Tarr, J.A. and Konvitz, J.A. (1987) Patterns in the Development of the Urban Infrastructure. In Gillette Jr., H. and Miller, Z.L. (eds), *American Urbanism. A Historiographical Review*. New York: Greenwood Press, pp. 196–225.

Tawa Lama-Rewal, S. (2007) La démocratie locale dans les métropoles indiennes. Les associations de résidents à New Delhi. *Transcontinentales. Sociétés, idéologies, système mondial*, 4: 131–144.

Terreblanche, S. (2004) La démocratie post-apartheid: un nouveau système élitiste? *Afrique contemporaine*, 210(2): 25–34.

Thelen, K. (2003) Comment les institutions évoluent: perspectives de l'analyse comparative historique. *L'Année de la régulation. Économie, Institutions, Pouvoirs, 7, 2003–2004*. Paris: Presses de Sciences Po., pp. 13–44.

Tournadre-Plancq, J. (2008) Protester dans l'Afrique du Sud post-apartheid. *Critique internationale*, 39: 139–160.

Triegaardt, J. (2007) Poverty and Inequality in South Africa: Policy Options and Consequences for Planning in an Emerging Democracy. Article written for the international conference, *Living on the Margins: Vulnerability, Social Exclusion and the State in the Informal Economy*. Cape Town: PLAAS/Isandla Institute/CPRC, 26–28 March.

Tsogas, G. (2011) Corporate Social Responsibility in Water Services: A Developing Country Perspective. In Wankel, C. and Malleck, S. (eds), *Ethical Models and Applications of Globalization. Cultural, Socio-political and Economic Perspectives*. Charlotte, NC: IAP, pp. 227–237.

Urquizo, C. (2009) *La Paz, ¿saco de aparapita o metrópoli andina?* La Paz: FES-ILDIS.

UIDSSMT Chandauli Project (n.d.) Document collected from the Jal Nigam office in Mughal Sarai.

Valfrey-Visser, B., Schaub-Jones, D., Collignon, B. and Chaponnière, E. (2006) *Access through Innovation: Expanding Water Service Delivery through Independent Network Providers. Considerations for Practitioners and Policymakers*. London: BPD Water and Sanitation.

Veolia Sustainable Development Report (2012) Available at: http://flipbooks.veolia.com/Veolia/radd2012/raddVFr2012/.

Verdeil, E., Féré, C. and Scherrer, F. (2009) De la rétroaction entre différenciation territoriale et modèle universel des services urbains en réseau: les enseignements du cas libanais. *Flux*, 75(1): 27–41.

Vickers, J. and Yarrow, G. (1989) *Privatization. An Economic Analysis*. Cambridge, MA, and London: The MIT Press.

Vincent, J.A. (2006) What Lies Beneath: The Inherent Dangers of the Central Arizona Groundwater Replenishment District. *Arizona State Law Journal*, 38: 857–880.

Vircoulon, T. (2004) La nouvelle Afrique du Sud. Une transformation à géométrie variable. *Études*, 401: 585–600.

Water and Sanitation Program (2010) *The Karnataka Urban Water Sector Improvement Project: 24×7 Water Supply is Achievable*. New Delhi: WSP, The World Bank, pp. 1–24.

Weis, B. (2005) Reflections on Urbanization in Water Infrastructure and Local Discourse in a Town in the Making. In Hust, E. and Mann, M. (eds), *Urbanization and Governance in India*. New Delhi: French Research Institutes in India and South Asia Institute/Manohar, pp. 151–175.

White, F. (2007) Afrique du Sud: les nouveaux mouvements sociaux de l'après-apartheid. In *État des résistances dans le Sud – 2007*. Louvain-la-Neuve and Paris: Centre tricontinental/Syllepse, coll. "Alternatives sud".

Wolfers, B. (2004) Privatization of the Berlin Waterworks – A Blueprint for Balancing Public and Private Interests? *Journal for European Environmental and Planning Law*, 1(2): 116–124.

Wolff, J. and de-Shalit, A. (2007) *Disadvantage*. New York: Oxford University Press.

World Bank (1994) *World Development Report 1994: Infrastructure for Development*. Washington, DC: The World Bank, Oxford University Press.

World Bank (2001) *Toolkit: A Guide for Hiring and Managing Advisors for Private Participation in Infrastructure. Executive Summary for Ministers and Senior Officials*. Washington, DC: PPIAF.

World Bank (2002) *India Poverty in India the Challenge of Uttar Pradesh*. Report No. 22323. Poverty Reduction and Economic Management Sector Unit, South Asia Region, The World Bank.

World Bank (2006) *Consumer Cooperatives: An Alternative Institutional Model for Delivery of Urban Water Supply and Sanitation Services?* Water Supply and Sanitation Working Notes no. 1, Infrastructure Network of the World Bank, ed. F. Ruiz-Mier and M. van Ginneken.

World Bank (2007) *India – Synthesis Study of Public Financial Management and Accountability in Urban Local Bodies*. New Delhi: Financial Management Unit, South Asia Region, World Bank.

Worster, D. (1986) *Rivers of Empire: Water, Aridity, and the Growth of the American West*. New York: Pantheon Books.

Wu, X. and Malaluan, N.A. (2008) A Tale of Two Concessionaires: A Natural Experiment of Water Privatisation in Metro Manila. *Urban Studies*, 45(1): 207–229.

Young, M.M. (2002) *Cooperative Infrastructures for Small Water Systems: A Case Study. Virginia Water Resources Research Center*. Special Report, SR 22–2002.

Zarbin, E. (1986) *Salt River Project. Four Steps Forward. 1902–1910*. Phoenix: COL Press.

Zérah, M.-H. (1999) *L'Accès à l'eau dans les villes indiennes*. Paris: Economica.

Zérah, M.-H. (2000) *Water: Unreliable Supply in India*. New Delhi: Manohar.

Zérah M.-H. (2003) Dix ans de libéralisation de l'économie indienne: les effets limités de la gouvernance dans le secteur de l'eau et de l'assainissement en ville. *Autrepart*, 27: 91–106.

Zérah, M.-H. (2006) Urban Water and Wastewater. In Rastogi, A. (ed.), *India Infrastructure Report 2006. Urban Infrastructure*. New Delhi: Oxford University Press, pp. 130–159.

Zérah, M.-H. (2009) Participatory Governance in Urban Management and the Shifting Geometry of Power in Mumbai. *Development and Change*, 40(5): 853–877.

Zibechi, R. (2005) Argentine: la fragile gouvernabilité. *La Jornada* (www.jornada.unam.mx), April (French translation by I.D. Reis and F. Lévêque, for RISAL).

Index

Page numbers in *italics* denote tables.

For Product Safety Concerns and Information please contact our EU representative GPSR@taylorandfrancis.com
Taylor & Francis Verlag GmbH, Kaufingerstraße 24, 80331 München, Germany

www.ingramcontent.com/pod-product-compliance
Ingram Content Group UK Ltd.
Pitfield, Milton Keynes, MK11 3LW, UK
UKHW011005180926
475701UK00004B/564

* 9 7 8 0 3 6 7 2 2 7 6 3 0 *